A Smart Energy System for Sustainable Buildings

Buildings

The Case of the Bernoulliborg

Faris Nizamic

Supported by the JoinEU-SEE Scholarship - Erasmus Mundus Partnerships Action 2, Ubbo Emmius Funds Scholarship, and the University of Groningen Green Mind Award.

ISBN: 978-90-367-8429-0 (book)
ISBN: 978-90-367-8430-6 (e-book)

Printed by CreateSpace, An Amazon.com Company
© 2016 Faris Nizamic

This thesis was completed using the thesis LATEXtemplate by Ando Emerencia, University of Groningen.
Proofreader: JoAnn van Seventer (joannvanseventer@home.nl).
Book cover idea: Faris Nizamic. *Book cover design*: Mersel Bujak and Eldin Herenda (http://newbusinessdepartment.com).

/ rijksuniversiteit
groningen

A Smart Energy System for Sustainable Buildings
The Case of the Bernoulliborg

Proefschrift

ter verkrijging van de graad van doctor aan de
Rijksuniversiteit Groningen
op gezag van de
rector magnificus Prof. dr. E. Sterken
en volgens besluit van het College voor Promoties.

De openbare verdediging zal plaatsvinden op

vrijdag 12 februari 2016 om 16.15 uur

door

Faris Nizamić

geboren op 6 december 1985
te Sarajevo, Bosnië en Herzegovina

Promotors
Prof. dr. M. Aiello
Prof. dr. A. Lazovik

Beoordelingscommissie
Prof. dr. E.L. Steg
Prof. dr. R. Baldoni
Prof. dr. A.J. Groen

To my beloved parents,
Fahrija and Rešad Nizamić

Contents

Acknowledgments

I would like to express my appreciation to those who played an important role in my life during the development of this dissertation.

This thesis is dedicated to my parents, Fahrija and Rešad Nizamić, who wished for me to achieve this degree more than anyone. As long as I live, I will be grateful for your unconditional love and limitless support in pursuing my endeavours. Unfortunately, my dad will not able to see this day, but I am so sure that he would be the most proud dad in this world, and that his smile and eyes would shine on the day of my defence. Dear Dad, I know that you would be proud and satisfied with what I have achieved. I am immensely grateful to my Mom, who lived through every single day with me, and who bravely took all the bitterness of being alone to make this achievement possible.

For my professional and personal development, I am most thankful to my academic fathers, Marco Aiello, Alexander Lazovik and Rix Groenboom.

Dear Marco, thank you for giving me an opportunity to pursue my PhD and for believing in me. Thank you for holding me to your high standards, teaching me to aim for quality, as well as planting in me some of your wisdom about the academic world and the world in general. Your generosity enabled me to follow most of my interests. Your open attitude, sincerity and friendship are something that changed me forever for the better, and I will never forget that.

Dear Alexander, thank you for accepting your role as my daily supervisor even though my research interests were slightly out of your field of expertise. What's more, thank you for your patience and for the numerous fruitful discussions we had! Even though sometimes I was not completely sure how to apply the feedback I received, it always gave me an additional push and new energy to carry on and to look for new ways to move forward.

Dear Rix, thank you for your early mentorship and numerous discussions about research, software testing, sustainability, Dutch culture, sports, and all other topics we share interest in. Thank you for pushing me to start quickly and to present my ideas at various prestigious conferences across the Netherlands. Thank you for making me feel at home while sharing nice food with you and your wonderful family. Mostly, thank you for your moral support, thinking-along and encouraging me to keep working.

I would also like to express my gratitude to the reading committee, namely Prof. dr. Linda Steg, Prof. dr. Roberto Baldoni and Prof. dr. Aard Groen, for assessing my thesis and for providing me with your valuable feedback. Your suggestions helped me improve my thesis and make it more complete.

I would like to thank my co-authors Rix Groenboom, Alexander Lazovik, Viktoriya Degeler, Tuan Anh Nguyen, Marco Aiello, Simon Harrer, and Guido Wirtz, as well Brian Setz, Ilche Georgievski, Ellen van der Werff, Berfu Unal and other colleagues whom I worked with on yet unpublished manuscripts. Special thanks go to Viktoriya for being my first PhD co-author. Through our joint work I understood how to write scientific papers. Furthermore, thanks for providing the inspiration for the introduction section of this thesis. Both you and Ilche shared very valuable knowledge about the Artificial Intelligence scheduling and planning techniques. Also, collaboration with Brian, Ilche and Tuan enriched my knowledge of service-oriented architecture design and implementation, subjects that comprise a very important part of this thesis.

Special thanks go to Ilche, Tuan and Brian.

Dear Ilche, thanks for being the first colleague to show me how to do paper reviews and one who helped me with a lot of practical things throughout my PhD. Your friendship, hard work and dedication are something that I admire and that I could only learn from. Moreover, speaking the Bosnian language with you helped me not to forget how to work using my mother tongue.

Dear Tuan, thanks for being a person with whom I shared most of the difficult tasks, and with whom I made very challenging and brave decisions. Together we made a great team and somehow it seemed that with you, achieving a new win is only matter of deciding to join the next challenge. I learned a lot from firefighting with you. I sincerely hope that we will use this synergy to make a lot of buildings more sustainable and that way make our contribution to this world.

Dear Brian, thanks for being my office mate, a person I could rely on, and the person who always remembered to speak Dutch to me when I started practicing it on a daily basis.

To all other current and former members of the Distributed Systems group, namely Saleem Anwar, Eirini Kaldeli, Pavel Bulanov, Ehsan Ullah Warriach, Ando Emerencia, Andrea Pagani, Doina Bucur, Fatimah Alsaif, Frank Blaauw, Heerko Groefsema, Azkario Rizky Pratama and Ang Sha, and Kerstin; I am very grateful for all the pleasant talks and discussions we had, as well as the time we spent together during various courses, group lunches, presentations, nice group trips and other fun activities.

I would like to thank to Ellen van der Werff and Marko Milovanovic from the Environmental Psychology Department from the University of Groningen for providing valuable inputs and revisions of several of the conducted surveys.

I would also like to express my gratitude to all the students I worked with or whom I supervised, for sharing an interest in the same topics and investing your time and energy to work toward joint solutions. Namely, I would like to thank: Diederik, Mattijs, Marcel, Sijmon, Brian, Ruurtjan, Rosario, Tuan, Bram, Nils, Niels, Bas, Jan, Jorrit, Rik, Maarten, Jeroen, Michel D., Thomas and Michel M. For me, you all were great sources of energy and inspiration, and I am grateful for that. Your results were always more than I could imagine at the moment when we started to work together.

I especially want to thank the employees of the University of Groningen for taking their valuable time to support us in implementation of the Green Mind Award projects. I express my gratitude to the sustainability manager Dick Jager, Sander Dijkstra, Yanike Sophie and members of the RUG Green Office; Marko Milovanovic, Maya Koekoek, Henri Hardieck, Edwin van Burum, Kor Smit, Ron ten Have, Jan Stalman and facility managers of the University of Groningen.

Special thanks to Peter Hartman, Hans Gaasendam, and Ronald Zwaagstra for their generous support in the Green Mind Award projects!

I would also like to thank Marille Zwaanenburg and Rein Smedinga for sending out all the surveys and spamming the staff members at my request. I would especially like to thank Mr. Dick Veldthuis and the FWN Board for giving permission for this research to be conducted within the Bernoulliborg building as well as to test prototypes of our start-up initiative (www.sustainablebuildings.nl) in its starting period.

An extra thanks is reserved for Gea Vellinga, Sarah Broersma, Willem Poterman, Maurits Alberda, Maureen van Veelen, and Marnix Pool, and other colleagues from the Venture Lab North and the Startup Fast Track from/with who I learned a lot about business.

Great appreciation goes to Esmee, Ineke, Desiree, and Helga for supporting me with administrative matters, and even more for making my day by sharing their wonderful laughs with me.

The greatest thanks goes to Janieta de Jong-Schlukebir, who was there for me for all four years (and longer), who had huge understanding of my situation, and who helped me in the moments when I needed that the most. As long as I live, I will always remember your unconditional support!

All the work would not have been done with satisfaction without the numerous fun moments I shared with my friends from Sarajevo and Groningen. Great thanks to my closest friends from Sarajevo: Ismar, Amel, Ervin, Aida, Mersel, Eldin, Lamija, Jasko, Selma, Anja, Adnan T., and their families who were there for me, both in happy and very tough moments. I am also grateful to all my family and relatives who believed in me and gave me warm words of encouragement, especially to my oldest aunt Aiša who often prayed for me to succeed.

For my Balkan friends in Groningen: Ena, Ivan, Edin, Adisa, Majo, Mirko, Igor, Maja and Marko, thank you for all the pleasurable moments when we reminded ourselves of our culture, customs, food and drinks. Special thanks go to Ivan for being my true mate, and a person who made my life in Groningen much more fun. I really enjoyed all our philosophical talks about life, love, sports, music, dance and many more.

Also, thanks to my Dutch friends from Groningen: Mirjam, Stephany, John, Rosanne, Erik, Petra, Gerard, and Maurits for sharing numerous happy moments and helping me relax when it was needed so I could get back and continue my intellectual work. And thanks to everyone else I have not managed to mention here and who shared nice moments with me.

Finally, biggest thanks goes to my dear Femke Smit for giving me all the warmth, care, and love that made this journey easier and much more enjoyable. From the first day, you made me feel at home and I will always be grateful for that. Your wonderful personality and great energy made my days colourful and filled them with lots of joy and happiness. I am also very grateful to your family for being so kind to me and for including me in Sinterklaas and other Dutch festivities.

Faris Nizamić
Groningen
January 18, 2016

Chapter 1

Introduction

I t is the first work day after a long vacation for Harvey, a user of our smart energy building. As Harvey approaches the building with his bicycle, a sensor at the entrance detects his arrival from his office key card and triggers the heating system to warm up his office to his preferred room temperature. After a few minutes, while Harvey uses the stairs to reach his office, his PC is also bootstrapping.

At the moment he approaches his office door a sensor detects his presence, unlocks the door and turns on the lights. From a room speaker Harvey hears the pleasant welcoming voice of his virtual secretary, Donna[1]: "Good morning, Harvey! It is 9AM and the room is ready for you. Room temperature is 20 degrees and your PC is on". While Harvey is hanging his jacket and taking items out of his bag, he hears more information about his schedule, missed phone calls and important emails to be answered. He is also being informed about total energy saving while on vacation and about the new energy saving goals of his department. As he wants to start with his work, he interrupts his virtual secretary by saying: "Thank you, Donna! That would be all for now".

At 11AM Harvey's colleague knocks on his door and enters to give Harvey an update about project developments while he was away. Harvey starts speaking with his colleague and after 3 minutes of PC inactivity his PC goes into sleep mode. As he was about to show some vacation photos to his colleague, he gets slightly annoyed by his PC going into sleep mode and wakes up the PC by touching the keyboard right away. That directly changes his PC sleep timeout from 3 to 5 minutes.

Now it is 1PM and Harvey leaves his office for lunch. The room sensors detect his absence and turn off the heating and put his PC in sleep mode. For one hour Harvey's office will not consume any energy. As Harvey arrives at the restaurant in the same building, lights are turned on only in the area where he sits, while other lights in neighbouring areas are dimmed. While he is enjoying

---

[1]The characters Donna and Harvey are borrowed from an American legal drama television series "Suits" (Korsh 2011).

lunch with his colleagues, he notices that lights above them also dim as the natural light level increases. As soon as he is finished with his lunch he passes the occupancy detection sensor at the staircase and the room preparation actions are triggered again. Only this time after he enters his office the lights do not turn on; a light sensor has detected enough sunlight coming from the outside. Furthermore, heating did not start 10 minutes but only 3 minutes before he entered the office; both outside and inside temperature sensors have detected a significant increase in temperature. This results in energy saving, about which Harvey is informed by his virtual secretary. Wanting to contribute further to the saving goal of his department, Harvey decides to cool himself down and asks his virtual secretary: "Donna, please notify Facility Management that my heating can be reduced by 20%. Thanks".

At 5.45PM Harvey finishes his work and leaves his office in a rush to get to his dining place on time. Sensors detect his absence and immediately turn off all energy consuming devices. Harvey's office again becomes energy neutral until the next working day.

This dissertation describes work contributing to the realization of such smart and sustainable office buildings. Most of the work in this dissertation was done as a part of the Green Mind Award 2012 project - *Sustainable Bernoulliborg*, which allowed for research ideas to be realized in an actual operating environment. The goal of the previous example is to present a vision of the future through a user-based story, as well as to illustrate a desired smart energy system that is not yet fully available on the market.

Different technologies contributing to the creation of such a system are appearing on the market. However, most solutions are expensive, difficult to install, and have to be operated by highly-skilled professionals. Furthermore, they are only partial solutions, such as a building management system covers only a few of the energy-consuming subsystems in a building (e.g., heating, cooling and ventilation). Moreover, a great amount of effort is needed to expand the solution from one to more locations.

In this thesis we therefore strive to produce solutions that can easily be adapted to different types of buildings, are easily extensible to cover more aspects of sustainability when they appear, and require minimum effort to be installed, configured and maintained in a new location. Moreover, we address economic affordability and user-friendliness, along with energy efficiency, as the main factors for both technical and business adoption.

1.1 Sustainability and ICT Systems

As defined by the U.S. Environmental Protection Agency: "Sustainability is based on a simple principle: Everything that we need for our survival and well-being depends, either directly or indirectly, on our natural environment. Sustainability creates and maintains the conditions under which humans and nature can exist in productive harmony, that permit fulfilling the social, economic and other requirements of present and future generations. Sustainability is important to making sure that we have and will continue to have, the water, materials, and resources to protect human health and our environment."

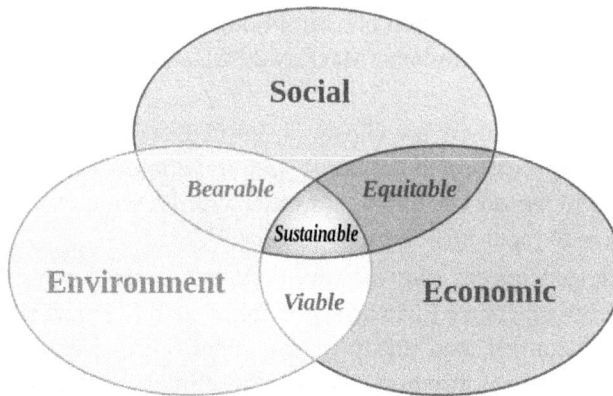

Figure 1.1: Venn diagram of sustainable development: at the confluence of three constituent parts

The 2005 World Summit on Social Development identified sustainable development goals, such as *economic development*, *social development* and *environmental protection*[2]. This view has been expressed visually using three overlapping ellipses, indicating that the three pillars of sustainability are not mutually exclusive and can be mutually reinforcing[3] (see Figure 1.1) (Cato 2009)[4].

To achieve true sustainability we need to balance economic, social and environmental sustainability factors in equal harmony[5]. These factors may be defined as: (1) Environmental Sustainability, (2) Economic Sustainability and (3) Social Sustainability. *Environmental sustainability* means that we are living within

---

[2]http://data.unaids.org/Topics/UniversalAccess/worldsummitoutcome_resolution_24oct2005_-en.pdf

[3]http://www.forestry.gov.uk/forestry/edik-59fmzf

[4]http://cmsdata.iucn.org/downloads/iucn_future_of_sustanability.pdf

[5]http://www.circularecology.com/sustainability-and-sustainable-development.html

the means of our natural resources. *Economic sustainability* requires that a business or country use its resources efficiently and responsibly so that it can operate in a sustainable manner to consistently produce an operational profit. Without operational profit a business cannot sustain its activities. Without acting responsibly and using its resources efficiently, a company will not be able to sustain its activities in the long term. *Social sustainability* is the ability of society, or any social system, to persistently achieve good social well-being. Taking these three pillars of sustainability further: if we achieve only two of the three pillars we end up with: Equitable, Bearable or Viable solutions.

Nowadays, the word sustainability is widely misused for political, commercial or other purposes. People are advised to undertake different environmental interventions so as to reduce their consumption. These manipulations are based on adjectives rather than numbers (MacKay 2008), and on estimations instead of real measurements.

Moreover, large numbers are chosen to impress, rather than to inform. As David J.C. MacKay explains in his book, "if everyone does a little [for sustainability], we'll achieve only a little". In other words, we as society as well as individuals have to focus on energy plans that add up.

Information systems can play an important role in raising awareness of and controlling energy efficiency in a variety of areas, such as smart cities and smart buildings, traffic control, and utility management (Pernici et al. 2012). In this work, we propose an information system whose main goal is to support optimization of use of resources within non-residential buildings, be they electricity, gas, water, or something else. Moreover, we strive to include more aspects of sustainability, such as an increase in waste recycling within buildings. This can have an impact on the environment and lead to less greenhouse gases emitted, less water or gas unnecessarily used and finally more waste recycled.

To implement such a system and make it in itself sustainable we have to take into account all three domains: economic, environmental, and social. A system has to be affordable and potentially generate economic savings, as well as to be socially acceptable in order to be implemented within an organization. On the one hand, if a system is too expensive, organizations will not be able to invest in it. Moreover, if a system does not break-even within an acceptable payback period, organizations will tend to postpone or reject the adoption of such a system and wait for solutions which bring more certainty or have shorter payback periods. On the other hand, if a system is not socially acceptable by its end-users, an organization will again not implement it or force its end-users to use it. In other words, to have a system that positively affects environmental aspects of an organization, the system must be both economically and socially sustainable.

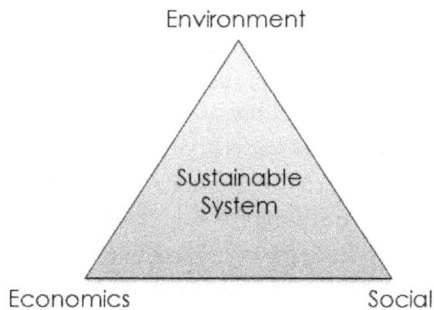

Figure 1.2: Sustainable system triangle

In line with the overlapping ellipses indicating the three pillars of sustainability, we define a sustainable system triangle (see Figure 1.2), which underlines three domains to be considered when providing a sustainable system. This triangle illustrates how in order to be adopted a proposed system should be within the acceptable limits of an organization (i.e., within triangle).

1.2   Non-residential Buildings

A building is regarded as a non-residential building when a minor part of the building is used for dwelling purposes[6]. Non-residential buildings include those with industrial, commercial, educational and health purposes, and any other buildings with non-residential purposes. Non-residential buildings also include buildings other than dwellings, involving fixtures, facilities and equipment that are integral parts of the structures and costs of site clearance and preparation. Historic monuments identified primarily as non-residential buildings are also included[7]. Examples of these are warehouses and industrial buildings, commercial buildings, buildings for public entertainment, hotels, restaurants, schools, universities, hospitals, elderly care institutions, etc.

Existing buildings are responsible for more than 40% of the world's total primary energy consumption (Howe 2010). It is estimated that at present such buildings contribute as much as one third of total global greenhouse gas emissions, primarily through the use of fossil fuels during their operational phase (UNEP 2009). Studies also show that on average 30% of the energy used in

---

[6]Source: OECD Glossary of statistical terms

[7]Source: Eurostat, "European System of Accounts - ESA 1995", Office for Official Publications of the European Communities, Luxembourg, 1996

non-residential buildings is wasted[8]. To tackle this issue, institutions such as the European Commission proposed several measures to increase efficiency at all stages of the energy chain, from generation to consumption. The EU's measures focus on those building sectors with the greatest potential for savings. However, on 28 June 2013, the European Commission published a report on the progress of member States towards Nearly Zero-Energy Buildings (NZEB), which are to become the norm for all new buildings in the EU by the end of 2020. The conclusion of the report is that too little progress has been made by the member states in their preparations, and that they will have to significantly step up their efforts to implement the requirements for NZEBs.

Today, many buildings are controlled using Building Management Systems (BMS). A building controlled by a BMS is often referred to as an intelligent building or smart building (Dragoicea et al. 2013). BMS are computer-based control systems installed in buildings to control and monitor the building's mechanical and electrical equipment. Current BMSs fail to reduce unnecessary energy consumption while preserving user comfort because, among other things, they are unable to cope with the dynamic changes caused by users' interactions with the environment (Nguyen and Aiello 2013).

Systems linked to a BMS typically represent 40% of a building's energy usage; if lighting is included, this number comes close to 70%. BMS systems are thus a critical component for managing energy demand. Improperly configured BMS systems are believed to account for 20% of building energy usage (Brambley 2005, Roth et al. 2002). Heating, ventilation, and air conditioning (HVAC), lighting, hot water, and electricity control are commonly seen as required functions of a BMS.

In addition, heating/cooling and lighting subsystems are usually controlled separately by isolated local feedback loops, thus valuable sensed information, e.g., occupancy patterns, temperature, light level, etc., are not shared and exploited by subsystems. Even more interestingly, computers in commercial buildings are not even considered as subsystems of BMSs.

1.3 Sustainability Initiatives in Built Environment

Let us consider three case studies of initiatives in building sustainability from The Netherlands. First, we present a sustainable state of the art office building. Here we highlight the costs of implementing such a building. Secondly, we present an example of an interesting data visualization ICT project and its challenges while trying to scale up to multiple buildings. Thirdly, we present a

---

[8]http://energy.gov/eere/buildings/about-commercial-buildings-integration-program

Figure 1.3: The Edge building, Amsterdam (NL) *(Photo: Ronald Tilleman)*

pioneering project aimed at solving a building energy consumption problem as an optimization problem, and we explain its advancements and drawbacks.

1.3.1 The Edge

At this point in time The Edge, as claimed by its creators, is regarded as the most sustainable building in the world. The Edge was awarded the highest score ever recorded by the Building Research Establishment (BRE). BRE is a world leading multi-disciplinary science centre with a mission to improve the built environment through research and knowledge generation. The Edge, located in Amsterdam (The Netherlands), achieved a BREEAM new construction certification of "Outstanding". BREEAM stands for BRE Environmental Assessment Method. BREEAM is the leading and most widely used environmental assessment method for buildings and communities[9]. By employing innovative smart technology the 40.000 square meter Grade A office building achieved a score of 98.36%.

According to The Edge website[10], by applying climate ceilings, LED lighting

---

[9]https://www.bre.co.uk
[10]http://www.the-edge.nl

and light over Ethernet, The Edge offers its users unprecedented user comfort. The climate ceilings provide radiant heat, comparable to floor heating. Light over Ethernet allows employees to use an application on their smart phones to regulate the climate and light preferences per workspace. An employee can adjust the light and temperature levels of his or her workplace. Light can be adjusted to a value between 0 and 500 lux and temperature in increments of 2 degrees Celsius from the default temperature of 21 degrees Celsius.

Moreover, on the façade and roof of the building rainwater is collected and reused. The ventilation system of the building recovers and reuses heat. A borehole underneath the building generates thermal energy through heat and cold storage underground. The building is shaped and oriented in such a way that the façades and atrium generate maximum daylight provision for the office floors and keep out the warmth of the sun at the same time. The south façade has solar panels on all parts that have no windows. The combination of all these sustainability measures in the building lead to a significant reduction in energy use and service costs. It is claimed that the development of The Edge will prevent about 42.000.000 kg CO_2 emission over next 10 years[11]. These estimations are based on the calculated energy performance of the building in comparison to the average Dutch office.

The building is also considered to be energy neutral. To achieve energy neutrality, the building owners partnered with the University of Amsterdam (UvA) and the Amsterdam University of Applied Sciences (HvA) to fit an area of their rooftops (4,100 square metres) with solar panels. Furthermore, an aquifer thermal energy storage (approximately 130 m. below the ground) generates all energy required for heating and cooling of the building.

According to the project manager[12], the success of this building lies in bringing together people responsible for offices, facility management and the ICT department. That enabled them to form a project team which could reuse all the benefits of concepts such as Smart Buildings, Internet of Things, Big Data and Data Analytics.

The proper infrastructure is crucial. In The Edge, the infrastructure is organized in such a way that energy consuming devices can be controlled on a level of a workplace or an office. This gives great opportunities for the optimization of office energy use. However, even though the building is considered to be energy neutral, the project manager mentioned that the calculations do not include

---

[11]http://www.stedenbouw.nl/amsterdam-the-edge

[12]On 30 June 2015, we visited The Edge to get closer look at the implementation. The guided tour was given by Mr Tim Sluiter from the Deloitte company, responsible for implementation of The Edge project.

plug load and devices used by the end-users. Therefore, there still may be some energy costs for the building.

All these advancements come at a price. According to information from the project manager, the cost of building The Edge from scratch amounted to about 210 million euros. This represents a significant investment. It is assumed that the payback period for this building will be longer than 10 years. Referring back to the pillars of sustainability, one may conclude that this project achieves only two out of three pillars, those of social and environmental sustainability, but not economic. According to this classification, The Edge project may be considered a bearable solution. However, for building owners, initiatives like this may also bring intangible returns of investment, such as improved public image, etc.

1.3.2 SMOG

The *Smart Metering Oldenburg Groningen*, in the following *SMOG*, was a project implemented within The North Sea Region Programme - Build with Care. The aim of the Build with Care project was to mainstream energy-efficient building design by raising the awareness and increasing the knowledge of the potential of energy savings[13]. The project ran between August 2008 and March 2012.

The aim of the SMOG project was to introduce energy consumption metering in the municipality of Groningen, The Netherlands. An application was developed for collecting and presenting energy usage data. By the end of the project 10 buildings, including office buildings, schools, a waste management site and a pump station, were connected to the energy data monitoring and displaying system.

The project was successful in several aspects. First, the prototype hardware and the software solutions were developed (see Figure 1.4) to enable collection of central energy consumption data[14]. The developed hardware solution, *the data collection box*, could be installed at locations with Internet access. Additionally, a knowledge base was created as a resource for further consultancy. During this project, algorithms were developed for automated pattern recognition within households. As one of the side effects, techniques to measure efficiency of heat pumps were developed and tested.

However, some aspects of this project were not successful. Even though an attempt was made to spread the results of the project, that initiative did not succeed to scale. It stayed at the level of a project; a product was never made and therefore it never reached the market. One of the reasons was that neither in-

---

[13]http://archive.northsearegion.eu/ivb/projects/details/&tid=74

[14]http://www.bits-chips.nl/artikel/geen-apparaat-blijft-verborgen-voor-slimme-meterkast.html

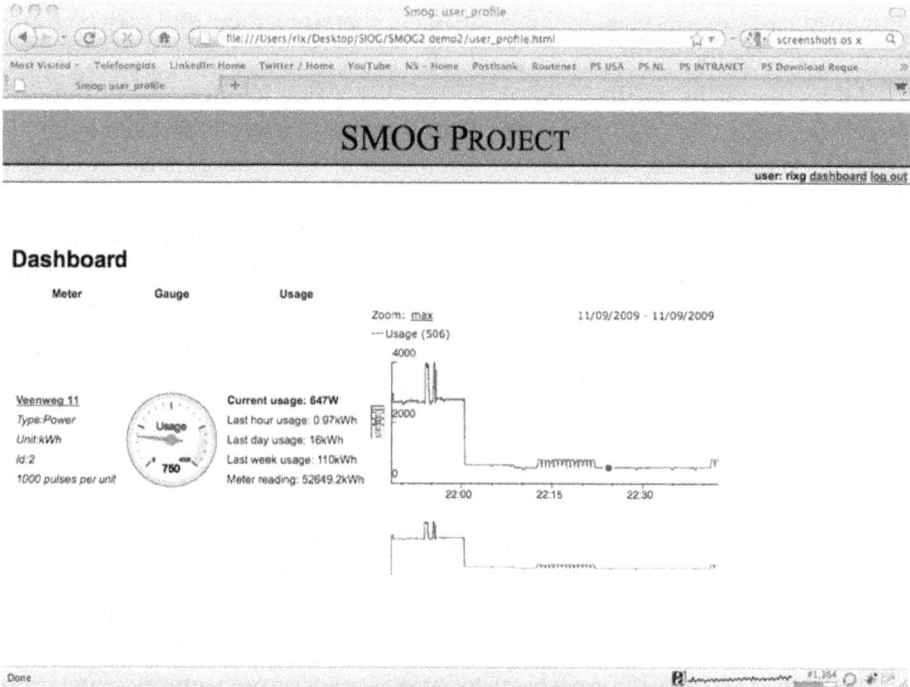

Figure 1.4: SMOG software front-end

vestors nor launching customers could be found to fund the next phase of development (Veenstra et al. 2008). That may also be due to several issues that affected the initial business model. At that time, the hardware (data collection box) was relatively expensive (750 Euro/piece). Moreover, electricity was cheap and potential savings were not sufficient to have a business case. The electricity price for non-household consumers in The Netherlands varies per quarter and per used consumption bandwidth (e.g., for quarter 1 in 2013, for consumers who have consumption from 500 to 2 000 MWh, the electricity price including transaction, delivery and network costs was 0.234 Eur/kWh)[15]. Furthermore, the venture had limited control over updates of both hardware and software, as it was initially outsourced to third parties. That led to very high costs for each new feature request. In addition, there were a lot of competing companies, which made entering the market even more difficult.

There were also some technical factors that affected the success of this initiative. The hardware was over-engineered and configuration of the data collection

---

[15]http://statline.cbs.nl/StatWeb/publication/?VW=T&DM=SLNL&PA=81309NED&LA=NL

box was complex. Furthermore, a relational database was used (i.e., MySQL) and that led to serious performance issues when scaling. The performance issues were due to sub-optimal database design and a bug in MySQL. Because of this bug a request to a page involving counting all of the data rows took more than 12 seconds to complete (Yumatov 2010). Rendering of graphs in the front-end was therefore unsatisfactory.

During an interview[16], one of the project leaders, Dr Rix Groenboom, stressed some of the lessons learned. First, it is important to understand what business problem is solved by using the developed technology. From that point, it is crucial to find a target market where these solutions can generate significant benefits. Next, it is important to understand from the beginning of a project that the solution will scale from tens to hundreds of buildings and to adjust software architecture accordingly so that the software can support the scaling process. Finally, the core business, in this case software and hardware development, should remain within the ownership of a venture so that solutions can be cost-effectively adjusted and improved during time. In the end, it should be relatively easy to maintain both the devices in the field as well as the software deployed on servers. These requirements can easily be translated to scalability, controllability, and maintainability requirements.

To conclude, this initiative failed to scale for the above-mentioned reasons. Moreover, the issue with this initiative, as with similar initiatives, is that they are implemented in the form of a project with a beginning and an end. Once the project is completed, the effort to bring the findings to the market is relatively high and requires a significant investment or a number of customers to finance it. Moreover, the technology has to be designed and implemented in such a way that it supports scalability.

1.3.3   GreenerBuildings

"GreenerBuildings[17] was a European FP7 project, which created a smart automated environment that combines automation for user satisfaction with energy-efficient environmental adaptation. As a part of the project, an intelligent office was constructed on the premises of the Eindhoven University of Technology, The Netherlands. The project allowed its users (i.e. people within a building) to establish and modify the rules of the building's behavior so that the system would automatically adapt to their needs by using the context information. The project featured advancements in many research areas, including wireless sensor net-

---

[16]First interview with Rix Groenboom, a business owner and the project leader of SMOG project took place in Groningen, The Netherlands, in May 2012

[17]http://www.greenerbuildings.eu

Living Lab PT313
Meeting Room

Number of people: 13
Outside temperature: 16C
Inside temperature: 20C
Activity: Presentation
Blinds height = 100%
Blinds angle = 0
HVAC 1: Off
HVAC 2: On
PMV: 0
Outside lux level: 2760 lux
Inside lux level: 500 lux

Figure 1.5: Living Lab PT313, Eindhoven University of Technology, The Netherlands

works, smart grids, activity recognition, thermo-fluid dynamics, etc." (Degeler 2014).

In the GreenerBuildings project, a full solution was implemented and we list the main contributions and strong parts of the solution as follows; first, a service-oriented software architecture was designed and implemented, and by the end of the project a fully functional system was demonstrated. Secondly, a service registry and dynamic service discovery were designed and implemented. Thirdly, the building energy consumption optimization problem was translated to and solved as an AI optimization problem. Finally, throughout the project it was concluded that an appropriate application domain for supporting scalability are non-residential (office) buildings.

After the project ended, support of the smart office in the Eindhoven University of Technology was discontinued. It seems that managing of the hardware infrastructure was not easy, requiring additional investment of time and effort to maintain the system. Furthermore, historical data although collected was largely ignored, e.g., the reasoning algorithms were based solely on streaming data, instead of combining the latter with historical data. To access streamline and historical data two different components were used.

Moreover, the scalability was main issue. Adding new locations, sensors, actuators, etc. was extremely difficult, from the points of both configuration and installation. The easiest way of scaling was to duplicate an existing working solution and to re-configure everything. However, that would eventually result in even larger efforts for maintenance. Next, there was inadequate communication with facility managers; in most cases the facility managers were not willing to cooperate. They did not see the benefits provided by the system but perceived it instead as an additional complexity. Finally, the project stayed at the level of one living lab due to the combination of the above-mentioned factors.

Looking back, one could conclude that the timing for this project was too early. Some of the technology at that time was not mature enough or not even available. Selected technology and hardware protocols did not provide added value or satisfactory quality. However, the architecture was well designed, although there were a couple of points to improve; for example, more emphasis should be given to how the infrastructure is managed and analytics should utilize available historical data. Moreover, more emphasis should be placed on the business perspective and economic benefits offered by the system. This would help to keep facility managers in the loop and include them in the project as well, resulting in better understanding of the system requirements.

1.4 Research Questions and Methodology

We define our research questions using the inspiration from the three previously described case studies. The first case study (The Edge) illustrated the important role of economic acceptability for organizations considering how to make their building(s) sustainable. The second case study (The SMOG) showed that lack of detailed requirement analysis can lead to inability for the system to scale, both technically and business-wise. The third case study (GreenerBuildings), along with the technical lessons learned, indicated that lack of good communication with facility managers and end-users may lead to lack of social acceptance of the system, resulting in low adoption. Finally, by analyzing all three case studies, one may observe that in the third case study energy saving is only partially tackled, and in the first case study it does not represent the main goal or is not advanced enough.

Therefore, to address all the identified issues and to develop a sustainable building, we have to look at the aspects of realizability, system efficiency, economic and social acceptability. These considerations are the basis of our research questions, to be discussed using Design Science Research Methodology (Hevner et al. 2004, Ken Peffers 2007). The research questions are as follows:

RQ1: Assuming that a smart energy system is realizable in an actual operating environment, how can real-time consumption data be feasibly obtained and what energy

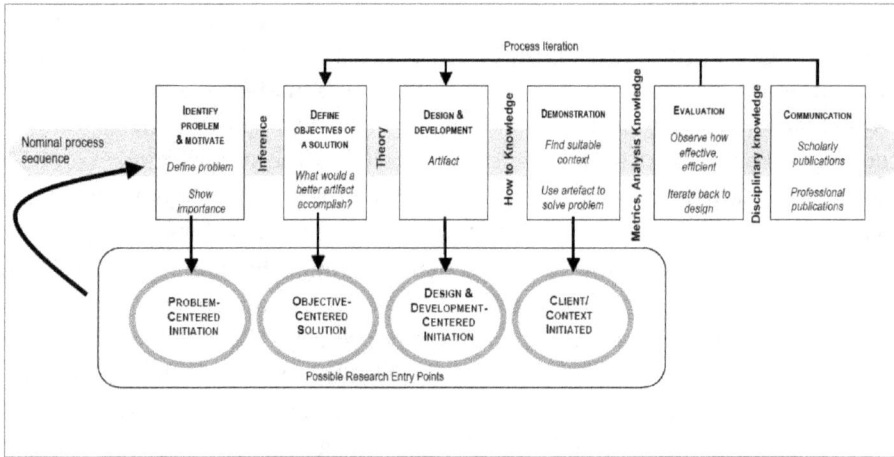

Figure 1.6: Design Science Research Methodology (DSRM) Process Model (Peffers et al., 2007)

consuming devices can be controlled using an ICT system? What should be the charac-teristics of an ICT system that supports sustainability measures in buildings?

RQ2: Can an ICT system increase the efficiency of resource use in a building, and if so, which techniques contribute to increase of efficiency while maintaining user comfort?

RQ3: Assuming that a smart energy system generates economic savings in a build-ing, which factors influence the economic acceptability of such a system? Moreover, which values of those influencing factors are acceptable and realistic?

RQ4: Is such a smart energy system acceptable for end-users of a building? If so, how important is social acceptability for such a system and which factors influence it?

To answer these research questions, the software and the hardware of a smart energy system should be fully implemented and deployed in a real building. Thereafter, measurement data needs to be collected and surveys conducted to evaluate all necessary aspects: efficiency, economic, social and other.

As previously stated, for our research we use the *Design Science Research Methodology* (DSRM). Design science creates and evaluates IT artefacts devel-oped to solve identified organizational problems. It involves a rigorous process to design artefacts to solve observed problems, to make research contributions, to evaluate the designs, and to communicate the results to appropriate audi-ences (Hevner et al. 2004).

"The Design Science process includes six steps: problem identification and motivation, definition of the objectives for a solution, design and development, demonstration, evaluation, and communication, as shown in Figure 1.6. *The first step* is problem identification and motivation. In this step, a specific research problem is defined and the value of the solution is justified. *The second step* is definition of the objectives for a solution. The objectives of a solution are inferred from the problem definition and knowledge of what is possible and feasible. The objectives can be quantitative, e.g., terms by which a desirable solution would be better than current ones, or qualitative, e.g., description of how a new artefact is expected to support solutions to problems not hitherto addressed. The objectives should be inferred rationally from the problem specification. *The third step* is design and development, or in other words, creation of the artefact. Conceptually, a design research artefact can be any designed object in which a research contribution is embedded in the design. This activity includes determining the desired functionality of the artefact and its architecture and then creating the actual artefact. *The fourth step* is demonstration. In this step, the use of the artefact to solve one or more instances of the problem is demonstrated. This could involve its use in experimentation, simulation, case study, proof, or any other appropriate activity. Resources required for the demonstration include effective knowledge of how to use the artefact to solve the problem. *The fifth step* is evaluation. In this step it is observed and measured how well the artefact supports a solution to the problem. This involves comparing the objectives of a solution to actual observed results from use of the artefact in the demonstration. Depending on the nature of the problem venue and the artefact, evaluation can take many forms. It could include such items as a comparison of the functionality of the artefact with the solution objectives from activity two above; objective quantitative performance measures such as budgets or items produced; and the results of satisfaction surveys, client feedback, or simulations. It could include quantifiable measures of system performance, such as response time or availability. Conceptually, such evaluation could include any appropriate empirical evidence or logical proof. At the end of this activity the researchers can decide whether to return to step three to try to improve the effectiveness of the artefact, or to continue on to communication and leave further improvement to subsequent projects. *The sixth and last step* is communication, in which the researchers communicate the problem and its importance, the artefact, its utility and novelty, the rigour of its design, and its effectiveness to researchers and other relevant audiences, such as practising professionals, when appropriate" (Ken Peffers 2007).

As our research is initiated by the observation of the problem (i.e., inefficient use of energy consuming devices in non-residential buildings), the research presented in this thesis is categorized as problem initiated design science. Therefore, we first propose a solution and define the objectives for a solution. Moreover, by gathering inputs from various stakeholders, we define the desired characteristics of the system. The next three parts of the thesis comprise the design, implementation and deployment process of the proposed system. Research contribution is embedded in the design as later described in the technological contributions section. The subsequent part of the thesis focuses on prototypes that lead to efficient use of resources by the developed system itself. It is important to acknowledge that the problem investigated in this thesis is not a new problem, but has been investigated by many researchers in the past decade. Therefore to explain how our proposed solution builds on and goes beyond the related work, in Chapter 8 we present the related work.

1.5 Our Case Study: The Bernoulliborg

The Bernoulli Building in a building of Faculty of Mathematics and Natural Science of the University in Groningen[18]. It is located at the Zernike complex in Groningen, The Netherlands.

Figure 1.7: The Bernoulliborg building, Groningen, The Netherlands

The building has a surface of 10.500 square meters accommodating 180 offices, 16 lecture rooms, 8 meeting rooms and 6 social corners for 350 staff members, and capacity for more than 500 students. The annual electricity consumption of the Bernoulliborg is between 1.350.000 kWh and 1.400.000 kWh. This is

---

[18]http://nl.wikipedia.org/wiki/Bernoulliborg

equivalent to the electricity consumption of approximately 420 average Dutch households.[19]

This building was chosen to be our case study for several reasons. First and foremost, the building is a workplace used by our research group, a well-known place to which our team members have everyday access. Secondly, universities and their buildings represent organizations that leave significant carbon footprints (between 100 tons to 1 million tons) (Berners-Lee 2010). Knowing this, the improvements made within this university could be replicated at more universities, thereby increasing the impact of this project. Furthermore, it is a relatively new (2008) and modern building with a building management system installed. Achieving optimizations in this building may prove that the same system could bring even higher savings in older buildings both with and without a building management system. Finally, by choosing this building we were able to participate in the Groningen University's competition - the Green Mind Award.

green
mind
award

Figure 1.8: The Green Mind Award competition, University of Groningen

The main conditions of the Green Mind Award competition are that: (1) The project idea should be a possible improvement in the field of sustainability within our buildings or our business operations; (2) The plan must be feasible in technical, practical, economic, legal and ethical terms; (3) The payback period must be no more than 10 years from the date of completion.

To carry out this research, in 2012 we were awarded the Green Mind Award (GMA) [20] for the project Sustainable Bernoulli building (1 of 62 submitted project proposals). The grant was awarded by the University of Groningen. The project had a limited budget of 100,000.00 Euros.

---

[19] According to statistics from https://www.wec-indicators.enerdata.eu/household-electricity-use.html, average annual electricity consumption per Dutch household for the period 2010-2013 is calculated to be 3265 kWh.

[20] http://www.rug.nl/about-us/who-are-we/sustainability/green-mind-award

The goal of the *Sustainable Bernoulliborg* project is to improve sustainability aspects at the University of Groningen, focusing specifically on the Bernoulliborg. During this project, several prototype solutions comprising a *smart energy system* were developed. We define a smart energy system as an ICT system that supports energy consumption reduction as well as integration of other sustainability interventions or systems.

The project included cooperation of a multidisciplinary team made up of researchers, project managers, facility managers, software developers, manufacturers and maintenance workers. In this project, among other things individual and common electricity consumption within the building was measured and the information presented to the building occupants by means of consumption displays. Moreover, lights in the restaurant and workstations in certain offices were automatically controlled. Finally, lighting use in the offices was optimized, and water consumption and the amount of general waste were reduced.

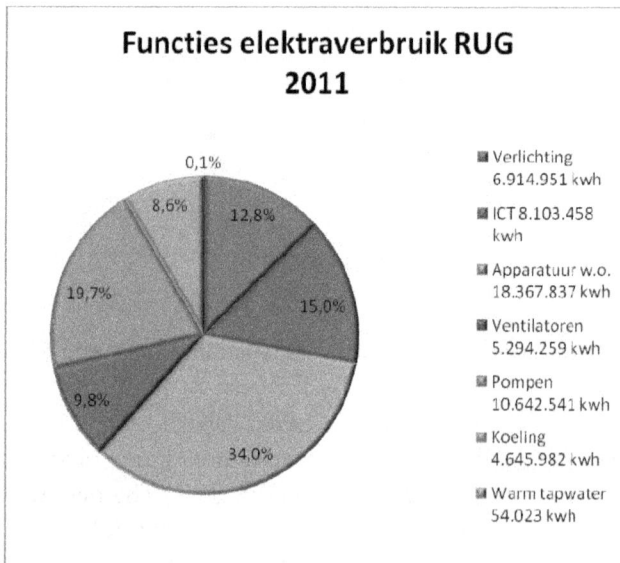

Functies elektraverbruik RUG 2011

- Verlichting 6.914.951 kwh
- ICT 8.103.458 kwh
- Apparatuur w.o. 18.367.837 kwh
- Ventilatoren 5.294.259 kwh
- Pompen 10.642.541 kwh
- Koeling 4.645.982 kwh
- Warm tapwater 54.023 kwh

Figure 1.9: University of Groningen total electricity consumption per subsystem for 2011

Figure 1.9, shows the division of energy consumption for the whole university for 2011. Figures presented were provided by the Energy Manager of the University of Groningen. Figure 1.9 shows the following percentages. Lighting accounts for 12.8%, ICT equipment for 15%, appliances for 34% and heating tap water 0.1%. Most of these consumers are not included as a part of the existing building management system. However, the ventilators (9.8%), pumps (19.7%) and cooling (8.6%) are in most cases included as a part of the current building management system.

However, it is important to understand that the total carbon footprint does not consist only of electricity consumption. To calculate the total environmental impact of an organization like a university, one should also include gas consumption, commuting, business travel, food, and everything that the university buys, right down to the smallest details (Berners-Lee 2010).

1.6 Thesis Contribution

In this thesis, we: (1) Conducted three case studies and explored related work, mainly with the purpose of understanding reasons for failure; (2) Gathered system requirements for designing and improvement of such systems; (3) Built on previous conclusions, improved architecture design and implementation of a smart energy system to support its function in an operating environment, as well as its scalability; (4) Demonstrated how such a smart energy system can be implemented in an actual operating environment; (5) Demonstrated how the utilization of computing resources can be optimized using AI techniques for scheduling and planning; (6) Evaluated savings from each implemented saving intervention; (7) Evaluated the economic and social acceptability of the system; and finally (8) Discussed factors that could increase acceptability of the system and therefore large scale roll-out. The technical contributions are explicitly mentioned and explained in more detail in the later sections.

1.7 Thesis Scope and Organization

The main goal of this dissertation is to illustrate the design, development and deployment of a smart energy system in a building, and to show how such a system is both realizable and also sustainable in its environmental, economic and social aspects.

The thesis starts by describing the context of sustainability, its three main pillars, and how it can be supported by an ICT system. Then the thesis describes non-residential buildings as significant contributors to resource inefficient usage as well as environmental pollution, representing the problem tackled in this thesis. Subsequently, the thesis demonstrates three initiatives attempting to increase sustainability aspects within a built environment, the first one by doing a heavy investment in design and development, the second one by introducing an ICT system for energy monitoring and management, and the third one by implementing an ICT system to support energy reductions through better user presence and activity recognition. These case studies show that for the success of a project all three main aspects of sustainability must be taken into account. Further, the thesis introduces the main case study of this thesis "The Sustainable Bernoulliborg" project. Finally, it introduces the research questions, research methodology and thesis contribution.

Chapter 2 proposes a design to solve the defined problem and presents the objectives of the solution. Then it proceeds with stakeholders and system requirements analysis. We collect inputs from the stakeholders, analyse them and present the full list of functional, non-functional and business requirements. We go on to present the technological contributions of this thesis. Then the chapter proceeds to discuss the design of the system, stating the design principles and presenting an overview of the architecture. This overview is followed by a presentation of the roles and dependencies of each system component. Finally, the chapter describes system communication and system operation. The work presented in this chapter was initially published in (Nizamic et al. 2014).

Chapter 3 presents a prototype implementation of a smart energy system called GreenMind system. We describe the implementation methodology. The main contribution of the chapter is the presentation of the implementation itself as well as the motivation behind the chosen technologies, as similarly described in (Nizamic et al. 2014). Moreover, we present the ways in which the system communicates, as well as the implemented data model and process of system integration. Chapter 3 also illustrates prototypes for regarding water and gas consumption reduction, waste management, and building inventory management.

In **Chapter 4**, we discuss how the implemented architecture is integrated and deployed in the Bernoulliborg. We describe living labs where the integrated solutions are deployed, as well as the users who are affected and who evaluated the deployed solutions. The logical units tackling separate building subsystems are grouped and presented with the deployed solutions.

In **Chapter 5**, we explain why it is important to optimize utilization of the underlying computing and storage infrastructure. Moreover, the chapter introduces two scenarios of cloud infrastructure optimization and goes into detail regarding two AI techniques: scheduling and planning. As will be seen in this chapter, these techniques are implemented and evaluated in the GreenMind system examples. With the examples presented, we show the feasibility of the approaches and how these reference implementations optimize resource usage in a cloud. The evaluation part of the chapter describes the performance of each technique used and compares the techniques with each other. We show that both techniques scale well for a typical size of the resource allocation problems considered in the thesis. More detail on material presented in this chapter can be found in (Nizamic et al. 2012).

In **Chapter 6**, we evaluate and discuss the results of the GreenMind prototype solutions, from environmental and economic perspectives, while in **Chapter 7** we present social acceptance considerations and results of related user acceptance surveys.

Chapter 8 presents scientific research related to energy efficient buildings, from computer science, psychological and economic perspectives. Furthermore, we provide details on related software architectures, as well as the application of related artificial intelligence algorithms. Finally, we present both already implemented and ongoing research projects, as well as several commercial products and services providing partial solutions to the problem of energy efficiency in buildings.

Finally, in **Chapter 9** we evaluate the overall system and show how the requirements are satisfied by the system. Furthermore, we give answers to the research questions, provide a summary of the chapters and reflection, and present general conclusions and suggestions for future directions.

Other work influencing this thesis was presented in (Nizamic 2013, Harrer et al. 2014, Nizamic et al. 2011). In (Nizamic 2013), we present the research ideas and concepts as to how service-oriented systems could be tested using simulation techniques. This knowledge was helpful during the software development process of our proposed system, as some delays were avoided by using the described technique. This work is enriched with a case study (Nizamic et al. 2011) including a simulation environment of a real business process. Finally, in (Harrer et al. 2014), we show how robustness of software (in our case BPEL engines) can be evaluated using different criteria.

Chapter 2

Designing a Smart Energy System

Just a few decades ago, our buildings were very simple. Electricity consuming devices were directly connected to the power grid and their usage mostly depended on needs and behaviour of users, turning devices on or off. Perhaps with an exception of a few relay timers installed here and there to turn off consumer devices when not used for a period of time, but not more than that. Unfortunately, a large portion of buildings are still like that.

A good candidate for building automation are non-residential buildings, more specifically, office buildings. Usually, they are not constantly occupied, as organizations, companies or institutions using those buildings operate based on their working hours. Moreover, during their working hours occupants spend some time outside their offices, for example having meetings or lunch brakes. This periodic absence of occupants gives an opportunity for energy savings. Moreover, non-residential buildings are often occupied by organizations that usually pay for their energy use, and therefore are interested and motivated to reduce their energy costs.

This potential is realized by large multi-national companies [1][2][3] which brought building automation systems to the market. Today, in many non-residential buildings automation systems can be found. Those systems mostly control heating, cooling and ventilation, and sometimes lighting and plug loads (Nguyen and Aiello 2013). Other consuming devices, such as computers and other office appliances are usually not included in those building automation systems.

With the latest technological improvements, computing and storage capacities, as well as sensing and actuating technology became more affordable. That represents a technology trigger that enabled better understanding of user behaviour and better control of consuming devices within buildings.

It is evident that there is a vital need to evaluate the building energy and comfort management systems in real-world situations (Nguyen and Aiello 2013).

---

[1]http://www.buildingtechnologies.siemens.com/
[2]http://buildingcontrols.honeywell.com/
[3]http://www.johnsoncontrols.com/

Therefore, we decided to take into account common state-of-the-art architectural patterns and best practices (Degeler 2014), as well as specific requirements of a system that has to be fully operational in real-world settings, and design the *GreenMind architecture*, an architecture for sustainable buildings. The key design goals of this architecture are performance, reliability, scalability and security.

One may wonder why an architecture of a smart energy system may help us to answer the research questions regarding energy efficiency, economic and social acceptability. A smart energy system is an information system which processes information on building usage, both from perspective of users and from perspective of energy consuming devices. This information system collects data, reasons about it and uses it for the control with a goal of optimization. By translating building management problem to optimization problem, we can use in our favor a range of already available optimization techniques (such as, AI planning and scheduling). This way, we can process previously unprocessed building information and that way provide answers to our research questions.

Using such an architecture, we are able to reduce energy consumption within buildings, provide and preserve occupant's satisfaction with their environment and ease the tasks of persons responsible for facility management. The architecture consists of components that are responsible for environment sensing, information gathering, decision making, and finally executing energy saving actions.

In this chapter, we propose a solution and state its objectives, collect system requirements, propose an architecture for smart energy system, define its components, their functionalities, and the way of communication between them. Finally, we describe how the system operates.

2.1 Proposed Solution and Objectives

As prescribed when following the Design Science Research Methodology (Section 1.4), we first propose a solution and define the objectives for that solution. To tackle the issue of inefficient use of resources (e.g., energy, water) in non-residential buildings, we propose a software architecture and implementation to support more efficient usage of resources within buildings.

The objectives of this system are to: (1) optimize resource consumption in non-residential buildings, (2) ease the tasks of building (facility) managers, so they can manage building(s) more efficiently, and (3) make occupants' environment more comfortable, so they accept and use the system.

The system should provide additional optimizations (e.g., energy savings) in comparison to current building management systems. It should be able to cope with the dynamic changes caused by users' interaction with building's environment. Moreover, it should include users input in the control loop, as they are important part of the system and one of the main contributors to resource consumption (as previously pointed in Chapter 1.2). Moreover, the system should provide user-friendly interfaces to enable interaction between a building, its managers, and its occupants. The system be easy to install and configure, to make it scalable from one room to a whole building, and then from one to more buildings. Finally, the system should be secure, without compromising privacy or physical security of the users.

2.2   System Requirements Analysis

To design a system that will contribute solving the defined problem and at the same time be acceptable by the stakeholders of the system, it is very important to collect the requirements. The process of collecting requirements is as follows. It starts with identification of the stakeholders. Once the stakeholders are known, data is collected from them. This data collection process is usually done using technique of interview. After interviews are performed, collected data is analyzed, categorized and presented in a form of system requirements.

The stakeholders of the system is anyone with a valid interest in the system. In other words, a stakeholder can be anyone who needs the system, benefits from the system, invests in the system, purchases the system, opposes the system, operates the system or uses the system.

In our case, there are three main types of stakeholders: energy managers, building or facility managers (further, facility managers) and the managing directors, mostly being building owners or business owners. The end-users of the system of the system are not included as the stakeholders they were initially not interested to contribute to gathering of requirements or were not technically sophisticated to do so. However, as we find them to be very important factor affecting a life of a building, we evaluate how they perceive the system and ask for their inputs for improvement.

The main stakeholders have different interests in the system. The energy managers have interest in the system as they have interest in optimizing energy consumption within their organizations and the system may support them to achieve their goals. The facility managers have interest in the system as they have duties of providing comfortable environment stimulating productivity of building occupants. Moreover, they are usually involved in energy saving interventions as those relate to environment they are managing and that way affect their clients, end-users of buildings. Finally, the third group are the managing directors, mostly being building owners or business owners who have direct interest in reducing energy costs. Building owners have interest in providing energy-optimal building, that will reduce energy costs for their tenants and that way make their facilities more attractive to their renters. Business owners may have multi-fold interest in the system. Firstly, they may want to decrease expenses of their operating costs. Secondly, they may want or need to reduce their CO_2 emissions as a part of governmental agreements (e.g., MJA3[4]). And finally, they may use the system to promote their commitment to sustainability and raise awareness of their employees, partners and visitors, building occupants. Besides these three defined groups, there may be more stakeholders involved. However, their influence to adoption of the system is quite limited and we do not include them in the following analysis.

To collect the requirements of such a system, we performed a cross-case analysis. Cross-case analysis is a research method that can mobilize knowledge from individual case studies (Khan and VanWynsberghe 2008). Our cross-case analysis consisted of an interview and/or a survey, involving all three most influential groups of the stakeholders: energy managers, facility managers and managing directors. This study was enriched with the literature review as well as our personal experiences and experiences of our research group of working almost a decade in this field.

Even though we discussed and collected the requirements with numerous managing directors being part of different organizations and companies, in this work we cannot address them by name. They range from university managers, hospital managers, energy coordinators, to owner of hotels, TV station energy providers and building construction companies.

In the following, we present the requirements that were most recurring and that showed to be important to interviewed energy and facility managers. Subsequently, we present the findings from our literature review. Finally, we add several requirements from our experience, experience of our partners, and perspective as the system provider.

2.2.1 A Cross-case Analysis

To gather requirements from facility managers and energy managers in a form of semi-structured interviews. We interviewed six facility managers and one energy manager of the University of Groningen. The facility managers are together responsible for 45 buildings, while the energy manager is responsible for 80 buildings in total. They are responsible for building operations as well as persons who are familiar both with the requirements of end-users (building

---

[4]http://www.rvo.nl/subsidies-regelingen/meerjarenafspraken-energie-efficiency

occupants) and organization management staff.

The requirements are gathered in a form of a semi-structured interviews, including both an interview and a survey (more in Appendix 2 - A Survey for Facility and Energy Managers). The interview consisted of five minutes for introduction to the system, ten minutes for three open questions, the rest of time for forty multiple choice questions, being at the same time a part of a survey. For further analysis, the audio recording of the interviews is performed. Each interview lasted for one hour. Most of the discussed features are graded by facility and energy managers with a grade on scale from 1 to 7. Average grade for each feature determined if a feature stays in the final set of requirements. The features which have average grade 80% from maximal (5.6) or higher are included in the final set. This excluded five identified features.

An Existing Literature Review

We additionally present findings from a literature review. We identified a similar study done in Germany[5]. The study reported on a requirements analysis for cost-effective energy metering system in commercial buildings. The study was conducted among managing directors, energy and facility managers form Germany. The objective of the study was to understand respective needs and demands, learn about challenges in commercial buildings and obtain use cases to derive functional and general system requirements. The study consisted of the following steps. First structured and in-depth interviews with 20 experts were conducted. These interviews led to documentation of use cases. Lastly, an online survey was used to prioritise use cases. The result of this study was that more than 50 use cases were collected. All of them relate to at least one of the following categories: energy, maintenance, automation, safety and compliance. As a result of this study, nine key system requirements are defined, namely: transparency, automatic provision of data and ubiquitous availability of reports and trends, analysis support, flexibility, real-time data, reliable data, accuracy, compliance, ease of use, and motivation.

A Business Owner Interview

We also include the findings from the interview of the project leader of SMOG project, as mentioned in Chapter 1.

First, the solution needs to be portable from one to tens, and later hundreds of buildings, to support scalability requirement. Therefore, the software needs

---

[5]http://2014.ict4s.org/files/2014/08/1-Understanding-Energy-A-Requirements-Analysis-for-Cost-Effective-Energy-Metering-System-in-Commercial-Buildings.pdf

to be designed that way to support the business scaling process. Second, as the solution needs to be adjusted frequently to support customer additional requirements, it should be controllable by our internal team that can provide agile and cost effective software and hardware updates. Finally, it should be easy to maintain devices in the field as well as software deployed on servers.

These requirements can easily be translated into the following requirements: portability, scalability, controllability, maintainability.

2.2.2   System Requirements

Taking into account all the inputs from cross-case analysis, literature review and business owner interview, we define the complete set of the requirements for design of the system. In the following, the complete list of the requirements is presented. The requirements are divided to functional (FR), non-functional (NFR) and business requirements (BR).

Requirement	**FR1 - Environment condition data collection (derived)**
Type	Functional
Description	In a building, the following factors should be measured: Light levels (for lighting control), Temperature (for heating and cooling control), Movement (for presence/absence detection), CO_2 levels (for the air conditioning control)
Goal	Understanding factual conditions of an environment as well as understanding how the environment is being used (occupancy).

Requirement	**FR2 - Consumption data collection and storage (derived)**
Type	Functional
Description	The system should be able to collect and store consumption data.
Goal	In order to support consumption monitoring, reporting and notification services, data about consumption has to be collected and stored.

Requirement	FR3 - Historic data collection (average grade: 6.43 out of 7)
Type	Functional
Description	Historic data about the activities and conditions in a building should be stored and reused for building operation optimization purposes.
Goal	Storing historic data for better understanding how environment is used over time, to understand trends, create profiles and predictions.

Requirement	FR4 - Monitoring WEB application for Managers (average grade: 6.71 out of 7)
Type	Functional
Description	A WEB application for facility and energy managers to MONITOR consumption within a building.
Goal	Managers should have quick access to monitoring system showing consumption of a building, so they can react fast to any noticed irregularities or unexpected consumption

Requirement	FR5 - Notification system for Managers (average grade: 6.29 out of 7)
Type	Functional
Description	A notifications system for facility and energy managers that informs if consumption exceeds certain planned or expected limits.
Goal	Facility and Energy managers should receive notifications or warnings when consumption of a building is more than expected, so they can react in time to bring consumption to normal or expected state, if possible.

Requirement	FR6 - Report generation for Managers (average grade: 6.57 out of 7)
Type	Functional
Description	Automated REPORT generation for facility and energy managers.
Goal	Managers should be able to generate reports from collected historical data, so they can understand the trends, relations between consumption and other influencing factors (e.g., weather, occupancy), so they can make more informed managerial decisions.

Requirement	FR7 - Monitoring personal consumption for building users (average grade: 5.71 out of 7)
Type	Functional
Description	User applications (e.g., mobile apps) for building users to MONITOR their PERSONAL energy consumption.
Goal	Building users should be able to monitor their personal consumption for the purpose of raising their awareness on energy consumption as well as to stimulate their energy conserving actions.

Requirement	FR8 - Monitoring overall consumption for building users (average grade: 5.71 out of 7)
Type	Functional
Description	User applications for building users to MONITOR OVERALL energy consumption of a building.
Goal	Building users should be able to monitor overall building consumption for the purpose of raising their awareness on energy consumption as well as to stimulate their energy conserving actions.

Requirement	FR9 - Control interface to Managers (average grade: 6.57 out of 7)
Type	Functional
Description	From the operational point of view, facility and energy managers must have ability to adjust how a building is being controlled.
Goal	As the Facility and Energy managers have a main role to manage buildings, for purpose of providing comfortable environment for its users, as well as to reduce energy consumption within their buildings, they should have (partial) ability and interface to control how energy consuming systems are being used within buildings.

Requirement	FR10 - Automated control of HVAC system (average grade: 6.57 out of 7)
Type	Functional
Description	The system should have ability to adjust HVAC system according to presence and/or activity of occupants inside of a building.
Goal	Reducing HVAC-related energy consumption by optimizing HVAC use.

Requirement	FR11 - Automated control of Lighting system (average grade: 6.71 out of 7)
Type	Functional
Description	The system should have ability to adjust LIGHTING according to light levels, presence and/or activity of people inside of a building, and other relevant parameters.
Goal	Reducing Lighting-related energy consumption by optimizing lighting use.

Requirement	FR12 - Automated control of Appliances (average grade: 6.00 out of 7)
Type	Functional
Description	The system should have ability to adjust APPLIANCES (PCs, printers, projectors, boilers) according to usage, presence and/or activity of people inside of a building.
Goal	Reducing Appliance-related energy consumption by optimizing use of appliances.

Requirement	NFR1 - Simplicity (average grade: 6.71 out of 7)
Type	Non-functional
Description	The system should have simple interfaces for users.
Goal	The system interface should be simple so users can comprehend graphics and get quick feedback on consumption.

Requirement	NFR2 - Installability (derived from business requirements)
Type	Non-functional
Description	The system should be easy to install.
Goal	The system should work on "plug and play" basis.

Requirement	NFR3 - Configurability (derived from business requirements)
Type	Non-functional
Description	The system should be easy to configure.
Goal	The system should be easily configurable by people with basic technical knowledge.

Requirement	NFR4 - Maintainability (average grade: 6.71 out of 7)
Type	Non-functional
Description	The system should be easy to maintain.
Goal	The system should be easy to maintain to support cost effectiveness requirement. The system should be maintainable from remote location through the Internet. Moreover, hardware faults or unexpected functioning of the system and/or parts of the system should be automatically detected and reported.

Requirement	NFR5 - Fault tolerance (average grade: 6.71 out of 7)
Type	Non-functional
Description	In case of unpredictable failures (e.g., power blackouts) the system should return to its normal working mode.
Goal	Ensure proper functionality of the system even after a non-graceful shut down.

Requirement	**NFR6 - High performance (average grade: 6.71 out of 7)**
Type	Non-functional
Description	The system should be able to react in the real time.
Goal	The system should react to users' commands and request in real time.

Requirement	**NFR7 - Performance (timeouts) (average grade: 6.29 out of 7)**
Type	Non-functional
Description	The system should use delays/timeouts to avoid too quick reactions to changes in environment.
Goal	The system should contain some delays before putting consuming devices into energy saving mode to minimize annoyance of users.

Requirement	**NFR8 - Privacy (average grade: 6.57 out of 7)**
Type	Non-functional
Description	Privacy of users should not be compromised within this system.
Goal	Privacy of users should be guarded, so users feel safe and unthreatened while using the system.

Requirement	**NFR9 - Scalability - a building (average grade: 6.29 out of 7)**
Type	Non-functional
Description	The system should be able to scale from the level of one room to the level a whole building.
Goal	System should be able to support operations in a whole building.

Requirement	**NFR10 - Scalability - more buildings (derived from business requirements)**
Type	Non-functional
Description	The system should be able to scale from one to hundreds of buildings.
Goal	Increasing scalability ensures sustainable business operations.

Requirement	**NFR11 - Portability (derived from scalability requirements)**
Type	Non-functional
Description	The solution needs to be portable from one to more buildings, to support scalability requirement.
Goal	System should be able to support portability from one to another location.

Requirement	**NFR12 - Security (average grade: 6.29 out of 7)**
Type	Non-functional
Description	The data collected by sensors (e.g., presence sensor) should not be accessible by unwanted third-party systems.
Goal	Reduce chances for eventual system misuse.

Requirement	**BR1 - Cost effectiveness (derived from business requirements)**
Type	Business
Description	The system should be cost effective. Payback time of the system should be within 5-7 years, in some cases 10-15 years.
Goal	System should pay itself back within time limits acceptable by potential clients.

Requirement	**BR2 - Controllability (derived from business requirements)**
Type	Business
Description	The system should be controllable by internal team that can provide quick and cost effective software and hardware updates/upgrades.
Goal	Support agile and cost effective software and hardware updates as the system needs to be frequently adjusted to support additional customer requirements.

2.3 Technological and Non-technological Contributions

As previously motivated in the case studies (Chapter 1.3), there are three main technological aspects that should be addressed, namely: managing of infrastructure, utilization of historical data, and scalability (i.e., configuration, installation). Non-technical aspects that should be addresses are also derived, especially lack of understanding of system requirements due to lack of communication with the main stakeholders (i.e. facility managers), as well as lack of inclusion of business perspective in design and development process. In the following, we explicitly state our technological and non-technological contribution in regard to our referent case studies.

First, in the present work more *importance is given to the way infrastructure is managed*. Now, it is easier to build and rebuild environment, making it more manageable. This is enabled by using micro-services as a software architecture style. The micro-services approach is a relatively new term in software architecture patterns. The micro-service architecture is an approach of developing an application as a set of small independent services. Each of the services is running in its own independent process. We use micro-services to support scalability and ease building, shipping and running of distributed applications. Moreover, we use cloud environments to deploy micro-services. This supports scalability goals as it enables single installation to be used by multiple parties (e.g., having single system to control hundreds of buildings). Even though in some cases this may be a single point of failure that affects availability of the service, it is still easier to maintain it than having to maintain multiple local installations.

Next, *analytics are done using historical data*. Whole architecture became more streamlined as historical data was promoted to play more important role. Stored sensor data is represented in time series. Time series are used to store *dynamic data*, that is being stored frequently, as often as one or more times per second and in large quantities. A time series is a sequence of data points, representing measurements made over a time interval. An example of this type of data are power consumption measurements. Using time series, scaling becomes easier as the constant and automated improvement of models using statistical analysis of time series and auto-regression becomes possible. Moreover, meta modeling is used for storage of *static data*. For instance, relation between structure of a building with devices (e.g., sensors and actuators) installed in a building and users of those devices.

Additional technical contribution that is motivated by the SMOG case study is more intensive use of Graphical User Interfaces (GUIs). We give *GUIs higher importance* as the main advantages of the proposed system come from human-machine interaction. GUIs are used to bring important information about consumption to the end-users, to raise their environmental awareness, as well as to help them actively and/or passively participate in environmental reductions. More importantly, we use GUIs to collect valuable feedback from the users. That feedback is used as input for control algorithms to improve the way energy consuming devices are controlled and that way increase effectiveness. This way, we bring users in the loop and take their reactions into account to create more comfortable environment.

Some technological contributions did not come as a direct motivation from previous projects and case studies, and represent new ideas. For instance, *to assess and adjust reasoning algorithms we use negative user feedback*. An example of negative feedback would be the case when the system triggers a PC to go to sleep mode, and after that action is executed, a user of the PC reacts right away by turning the PC back on again. This indicates that the user was not satisfied with the control action and it is useful to correct the control algorithm. Negative user feedback is used more than positive feedback to measure success of reasoning. Negative user feedback also helps with learning from reasoning mistakes and further fine tuning of reasoning algorithms. Reasoning algorithms contain events which may include this feedback. Moreover, user feedback became a measure of success of reasoning algorithms.

Two non-technical contributions are the following. First, *facility managers are a part of the development team*. This fact made them aware that ICT can support them in being more productive. That resulted in willingness and improved cooperation with facility managers, what made gathering of the requirements and fine tuning the whole system easier. And second, the business perspective of the project gained on importance and *business requirements are seriously considered during the design and development process*.

2.4   The GreenMind Architecture

Towards the satisfaction of the identified system requirements, we propose an architecture of a smart energy system that is capable of integrating more building consumer subsystems and devices, and can provide a continuous near optimal energy control.

This architecture takes as its inputs: (a) user activities, such as working with PC, being present or absent, (b) appliance statuses, e.g., PCs are idle for a certain amount of time, and (c) environment information (natural light intensity, outdoor temperature, etc.). The system adjusts the environment to preserve user comfort and to save energy. The proposed architecture goes from the physical level of consumption measurement, live environment sensing, up to reasoning and controlling. Furthermore, the system also provides user interfaces that bring consumption information at the building level, but also at the personal level.

2.4.1 Design Principles

As described in (Tanenbaum and Steen 2006), type of the system described by the requirements is categorized as a distributed pervasive system. Pervasive computing (also called ubiquitous computing) is defined[6] as is the growing trend towards embedding microprocessors in everyday objects so they can communicate information. The words pervasive and ubiquitous mean "existing everywhere." Pervasive computing devices are completely connected and constantly available. A distributed pervasive system can become a part of our surroundings by utilizing services provided by sensor networks. As the identified requirements (especially, non-functional ones) comply with the main characteristics of distributed systems and service-oriented architecture (i.e., fault tolerance, scalability, openness, transparency), we decided to follow the founding principles (Tanenbaum and Steen 2006) of these paradigms in design of the proposed architecture.

The organization of the system consists of multiple subcomponents. The components work together over a network to reach a common energy saving goal, sharing useful information with the other components so that both the raw sensor data and other processed information is accessible to each component. Moreover, the components should be *abstract*, in other words they should hide inner logic from the outside world.

To make the management of these components easier, the system must be *modular* and ready for future extensions, the components must be *loosely coupled* and the communication must be *asynchronous*. For all mentioned reasons, we decided to design the system using the architectural pattern of *Service-Oriented Architecture (SOA)* as it encompasses all necessary principles (e.g., loose coupling, abstraction, encapsulation, etc.). SOA is an architectural style that supports service-orientation. Service-orientation is a way of thinking in terms of services and service-based development and the outcomes of services[7].

It is envisioned that over time new sensors and actuators can be added to the system, as well as new components that provide better or completely new functionality, so that the system keeps improving. Through the loose coupling

---

[6]http://searchnetworking.techtarget.com/definition/pervasive-computing
[7]http://www.opengroup.org/soa/source-book/soa/soa.htm#soa_definition

of the individual components the system scales better, it adapts more easily to new components, and it is easier to manage than a conventional building management system. By using a dynamic combination of hardware, that integrates with the existing building design, and dedicated software, the system has the potential to save energy at a relatively low cost without the loss of comfort and even with the potential to increase comfort.

2.4.2 Architecture Overview

The architecture of the GreenMind system is shown in Figure 2.1. The architecture consists of four layers: Physical, Utility, Reasoning and User layer. Each layer consists of more components and is responsible to serve and to provide functionality for a neighboring (upper or lower) layer. In the following, we describe characteristics of each component in short as well as their dependencies on other components. Moreover, we map the functional requirements to specific components, where appropriate. Non-functional components are addressed in the Implementation Chapter, where more details on each component are also given.

At the bottom lays the Physical layer that contains all devices connected to the Gateway. The Physical layer also contains Hardware Interfaces (HIs) and Software Interfaces (SIs). Hardware Interfaces (HIs) interface with real devices for the purpose of monitoring and controlling (*FR1, FR2, FR3*). Software Interfaces (SIs), such as a software agent that runs on clients machines (i.e., PCs), works both as a sensor for monitoring and as an actuator for controlling workstations.

The Utility layer consists of five components: Context (*FR1*), Consumption Measurement (*FR2*), Databases (*FR2, FR3*), Orchestrator, and Sleep Management component. The data previously processed by the Gateway is very important for the Context component to reason and provide a complete and consistent view of the environment. This essential view is stored in the Databases together with consumption data from the Consumption Measurement component. Also, the crucial view about the environment is provided to the upper – Reasoning layer component, the Decision Making. The Decision Making makes decisions on how to control the devices (*FR10, FR11, FR12*). The control decisions made by the Decision Making are sent again to the lower – Utility layer, to the Orchestrator. The Orchestrator translates them into proper commands distributed to either the Gateway that takes care of actuation or the Sleep Management (SM) component that takes care of control of workstations through interaction with the Software Interface (e.g., Sleep Management software agent). The uppermost layer of the architecture, the User Layer, contains the components that are responsible for

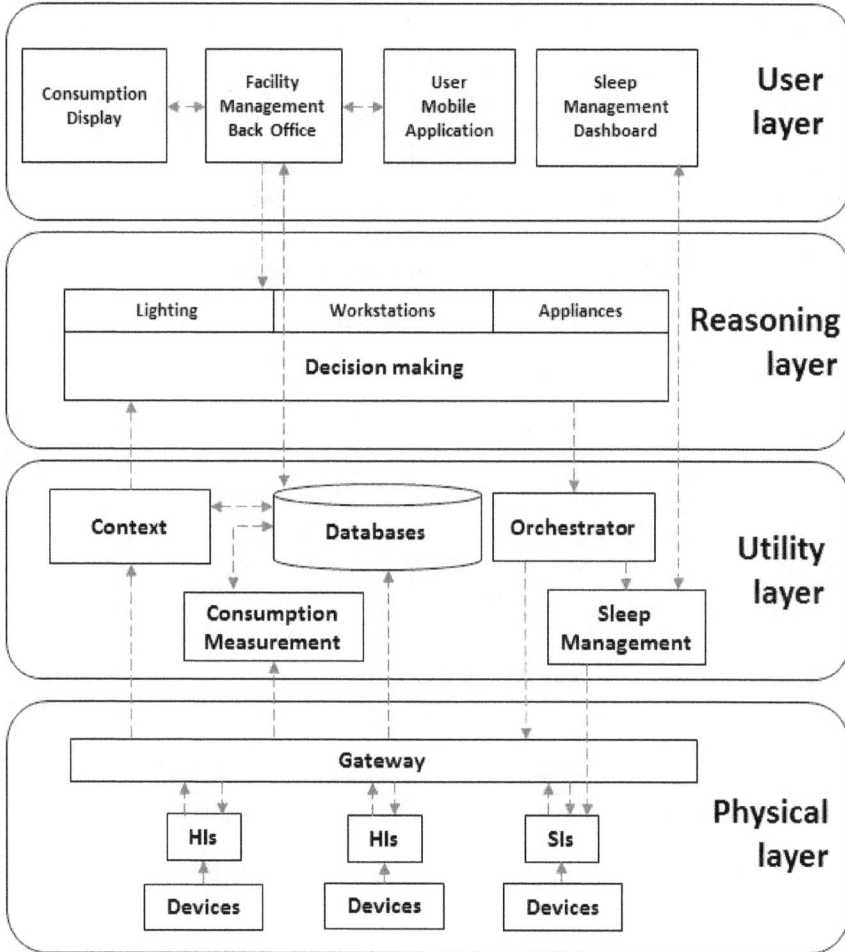

Figure 2.1: GreenMind system architecture

delivering the consumption information to occupants of a building as well as for monitoring, configuration and control of the system by the user (*FR4, FR5, FR6, FR7, FR8, FR9*).

Let us now describe each layer and component in detail.

2.4.3   Physical Layer

The Physical Layer represents the connection with the operation environment and actual building energy consuming devices. In office buildings, consuming devices are usually lights, heating, ventilation, and air conditioning, workstations and plug appliances. Other devices we consider to be on physical layer are sensors and actuators. They form sensor networks and provide the basic infrastructure for gathering raw contextual data. To gather data from the physical environment and to control the consuming devices, we connect them to Hardware Interfaces (HIs) and/or Software Interfaces (SIs). HIs and SIs communicate further with the Gateway responsible for protocols translation and data unification.

Hardware Interfaces (HIs) and Software Interfaces (SIs)

The Hardware Interfaces (HIs) represent hardware devices that enable data collection and interaction with consuming devices, sensors and actuators. The HIs use different protocols to communicate with sensor networks. Through HIs, we collect raw contextual data. For example, from light and temperature sensors we collect information about the ambient context, from passive-infrared (PIR) and pressure sensors we collect information on users' presence and activity, while from electricity measuring plugs we gather information about device use and device electricity consumption.

The Software Interfaces (SIs) offer a facility for controlling physical devices using software implementation. The SIs come in the form of client software implemented on end-user devices that can be controlled programmatically. For example, a client application can be installed on a workstation to monitor power management events and workstation activity while listening to and executing commands to control the state of a workstation (on, off, sleep mode).

Gateway

The Gateway represents a node that is responsible for interfacing with sensor networks that use different protocols. It serves as a protocol translator as well as fault isolator. The gateway provides system interoperability among different sensor networks. The environment is being interpreted through information gathered by sensors. The more sensors are deployed, the more precise information can be deduced.

The diversity of sensors implies that the system should be able to address the heterogeneity of sensor networks. Thus, the Gateway is capable of reading information from different networks and join data read from them into only one standardized message with a specific format, realizing the first step of data processing. It also provides an interface for upper layer components. In addition, the Gateway encapsulates the complexity of sensor networks and makes components independent from each other.

2.4.4 Utility Layer

The Utility Layer is responsible for facilitating the main functionalities of the system, such as environment sensing, data storage and processing, and environment actuation. It consists of five components: Context, Consumption Measurements, Database, Sleep Management and Orchestrator.

Context Component

The Context takes as its input the sensor data processed by the Gateway. The Context component has two subcomponents: (1) *ambient context* and (2) *activity recognition*. The ambient context takes care of ambient information, such as natural light intensity, temperature, humidity, etc.

For user activity recognition, we apply an activity recognition approach to derive high-level activities from simple sensor data. We adopt the activity recognition solution presented in (Nguyen, Raspitzu and Aiello 2013). For more detail about the implementation of our context component, we should refer to (Nguyen, Degeler, Contarino, Lazovik, Bucur and Aiello 2013).

Consumption Measurement

The Consumption Measurement component is responsible for measuring consumption at the room or even device level, for instance, electricity consumption by an occupant. Besides occupants' devices (e.g., PC, fixed line phone, mobile phone charger, water cooker), this component may be used to measure office light consumption. That way, it is possible to determine total consumption per room or per user. This enables keeping of a room or a user consumption history, for provision of feedback on the energy consumption.

This component enables fine-grained sub-metering, at the occupant level, which gives additional possibilities for data analysis and data presentation to the end-users (i.e., occupants). Knowing consumption on this level gives an opportunity to understand usage patterns which may in future support with usage prediction as well as organization of some motivating energy conservation competitions among occupants (Jilings and Heitmeijer 2013, Dekker and Brandsma 2015).

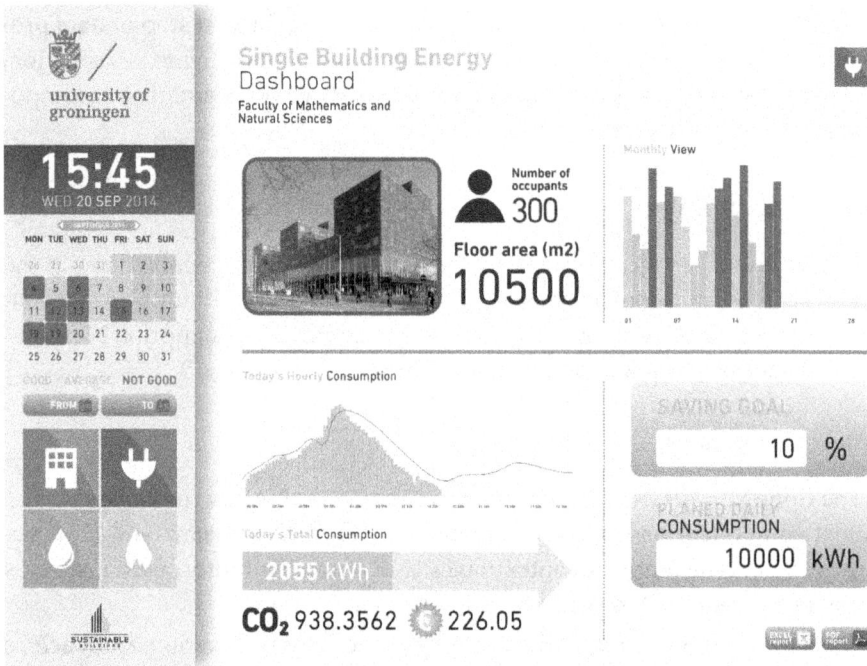

Figure 2.2: Graphical design of the Consumption Display 2.0 - Energy Dashboard (Design by: Eldin Herenda, NewBusinessDepartment.com)

Databases

The database is necessary to store and retrieve the data (e.g., the sensor measurements, or buildings structural data) for the components of the system. The data can be stored in relational databases using tables, in a graph database, or both. The primary difference is that in a graph database the relationships are stored at an individual record level, whereas in a relational database the structure is defined at a higher level (i.e., table definitions). One of the main advantages is that graph databases provides index-free adjacency (i.e., graph traversals can be performed with no index lookups) that in case of a large amount of data points can lead to a high performance.

Each sensor measurement contains relationships to other data such as the floor, building, time and date, and room or user. As a large amount (e.g., 1 million records) of sensor measurements with a traditional database caused severe performance issues (e.g., retrieve the total energy consumption for a certain hour on a specific date), we use a combination of NoSQL database and graph database. NoSQL database delivers faster performance for systems involving time series or Big Data (such as data coming from thousands of sensors), while graph database has advantages for maintenance of building structure and relationship between nodes (e.g., users, rooms, devices).

Sleep Management

The purpose of the Sleep Management (SM) solution is to minimize the energy usage of workstations (i.e., desktop PCs, laptops). The way SM achieves this goal is by taking control of the process for putting a workstation into sleep mode when no activity is detected. The SM also provides the administrators with important information with regards to the activity history of a workstation and whether a workstation should be in sleep mode or not. The SM is a server managing data received from all clients and listening to requests from other components, forwarding these requests to the clients. More technical details on SM component can be found in (Setz and Pul 2013, Setz 2015).

Orchestrator

The Orchestrator component acts as a buffer between the reasoning layer and the physical layer. The Orchestrator takes care of execution of control commands by sending the commands at the intended time, as well as the order of execution. To limit chance of the orchestrator component becoming a bottleneck, the orchestrator is multi-threaded and implemented in such a way that multiple instances can execute simultaneously.

The orchestrator communicates with the Sleep Management and the Gateway components, depending on the instructions received from the upper components. The SM handles the requests from the orchestrator to change the state of workstations and the gateway handles the state change requests for the other devices.

2.4.5 Reasoning Layer

The Reasoning Layer consists of the Decision Making component. This layer is represents the brain of the system and is responsible processing of sensor data and provision of energy saving actions or commands to be executed by the lower layers.

Decision Making

The Decision Making component uses all the available data and functionality from the lower layers to make decisions about the desired state of devices. The

Decision Making component is responsible for providing the system with the ability to automate the sensor and actuator behavior, in turn influencing energy consumption.

Given the context information and the recognized activities, AI planning (Georgievski and Aiello 2015) is adopted to compose a plan, that is, a sequence of device operations. Subsequently, plans are executed by using a network of actuators. The Decision Making component takes care of controlling lighting systems, but it also invokes the Sleep Management component for controlling workstations (i.e., PCs) in order to reduce energy consumption while not hindering a user.

For example, the Decision Making component receives data from sensors (e.g., light/lux and movement/PIR sensor) previously processed by the Gateway and Context component inputs from. Then, it uses a AI planning technique to compute desired actions and provide set of actions to the Orchestrator which further sends them to the Gateway to actually execute these actions on the Physical layer (e.g., turn off or turn on the lights). Similarly, PCs are controlled by sensing the activity of users (e.g., keyboard or mouse activity) and then closing the loop from the Orchestrator through the Sleep Management component, which uses the Software Interfaces to send actual commands to PCs (e.g., put them in sleep mode, turn off, etc.). The same logic can be applied for other energy consuming systems, with the goal of optimizing utilization of devices and that way providing energy (and other) savings.

2.4.6   User Layer

This layer comprises interfaces for both facility managers and end-users. Through these interfaces users can get information about the state of the environment being sensed (e.g., occupancy, ambient or device state information), as well as consumption information (e.g, power, gas, water, or other consumption). Moreover, users can use these interfaces to provide preferences or feedback regarding state of the environment (e.g., temperature and light level in a room), to control devices and systems manually, or to give input and/or enable automated control.

Consumption Display

To raise awareness among occupants and decrease energy consumption by changing their behaviour, it is important to bring users in the loop by providing them information about the energy consumption. Depending on the implementation, the display may present the information on different levels, from a overall building consumption to individual user or even device level information. The graphs may be presented with varying level of granularity, more specifically in daily, weekly and monthly view. Also, meaningful information such as a price of consumed energy, a saving goal of a group of occupants or a whole organisation, as well as an amount of emitted CO_2 may presented on a display. Figure 2.2 shows a graphical design of the Consumption Display v2.0 (Energy Dashboard).

Facility Management Back Office

The Facility Management Back Office component is responsible for providing an administrative panel for configuration, logging and monitoring. Additionally, it is responsible for providing Mobile Application clients with an API for authentication, authorization and data retrieval as well as pushing notifications (e.g. progress updates, achievements).

This component depends on the databases for retrieving relevant information concerning both personal energy consumption as well as the building structure and data on total consumption of a building. Through this component, an administrator of the system can monitor the health of the system, verify logs for any unusual system behaviour, import structural data about a building as well as map a mobile application user to a corresponding measuring device(s).

User Mobile Application

The User Mobile Application provides users with personalized usage statistics and overall building statistics (Meiboom 2013). The Mobile Application notifies users of progress on energy goals. It enables users to keep track of consumed energy, to set or inherit a saving goal for a certain period (e.g., week, month), as well as to participate in saving campaigns managed by building or facility managers, etc. This component depends on the Facility Management Back Office component for retrieving data.

The User Mobile Application component brings data about consumption closer to users and keeps them aware of the energy they consume. If widely used in an organization, this component can be also used as an additional communication channel with building occupants to provide important notifications, alerts, etc.

Sleep Management Dashboard

Using the Sleep Management Dashboard (SMD) graphical user interface, we see the workstation (e.g., PCs) activity of users and have ability to send workstations to sleep mode, wake them up, turn them off or on; or simply view when a

PC was off, working or in sleep mode. Some additional options, such as calculating of savings per PC are possible.

Having this component, we include workstations as a part of the overall energy management and building automation system. Moreover, this component can be potentially reused for presence detection in buildings.

2.4.7 System Communication

System communication is based on message queues, software-engineering components used for interprocess communication. These are used by the components of the system to communicate with each other, with or without knowing the physical location of other components. Message queues can significantly simplify the implementation of the separate components and also improve performance, scalability and reliability.

Another advantage of using a message queue is that the sender and receiver do not need to communicate with the message queue at the same time as the messages are stored onto the queue until the recipient receives the message. As a result, the communication between the components is asynchronous.

We use a message broker to reduce coupling between low-level components (i.e., sensor networks and Gateways) and the high-level ones (i.e., Context), which makes it possible for the components to run independently in a distributed manner.

2.4.8 System Operation

The typical operation cycle is as follows: 1) the Gateway sends messages to the Context based on the changed values from their sensors and actuators, 2) the Context component processes these messages to create new variables and sends the variables values to the Decision Making component, 3) the Decision Making executes planning based on the received values, 4) messages that require action are then sent from the Decision Making to the Orchestrator, 5) the Orchestrator then sends commands to either the Sleep Management or the Gateway, depending on the received message, and 6) the Sleep Management or the Gateway then execute the requested action (e.g., turn off a device or put a workstation to the sleep mode), and then the cycle repeats. The components do not wait until a full cycle is completed (e.g., a change in the sensor value can occur at any moment). Closing the full circle from sensing the changes in the environment to control of the end-devices provides the feature of automated control. The end-devices can be any energy consuming device that we can interface with, for example lighting, appliances, HVAC, as described with functional requirements FR10, FR11, and FR12.

To give examples of system operation, we present here two use cases, one for lighting and one for workstations. First, let us consider the lighting control (illustrated in Figure 2.3). In offices, a number of light sensors and presence sensors are deployed. Those sensors continuously sense the environment and send their measured values to the Gateway. The Gateway unifies messages coming from different sensors and sends them to the Context component. In the Context component, raw sensor data is transformed into new variable values (e.g., *Light level=high*, *Presence=false*), that are understandable by the Decision Making component. The Decision Making executes planning based on received values (e.g., *high, false*) and if action is required (e.g., *turn lamps OFF*), it sends set of actions to the Orchestrator. In this case, when the light level is high and no one is present, lamps should be turned off, and that is what Orchestrator is instructed to do. Finally, the Orchestrator sends commands to the Gateway to execute the requested action, namely turn particular lamps off.

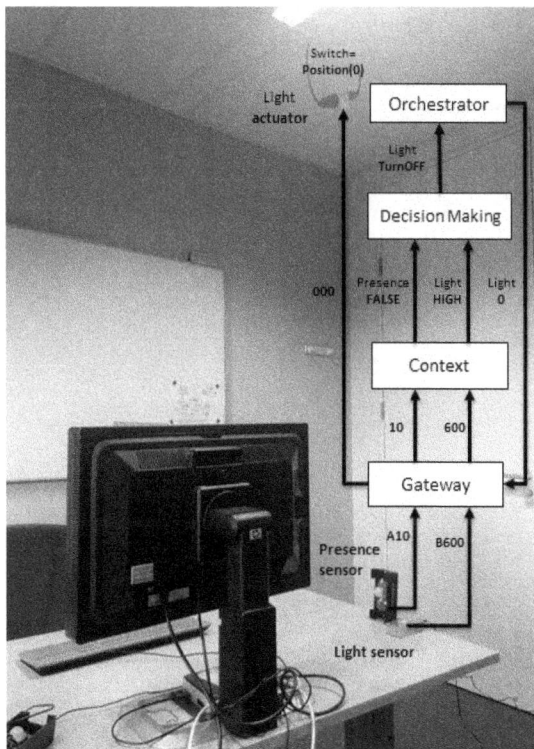

Figure 2.3: Illustration of the system operation: case of lighting control

The second use case concerns workstations control. Inputs are now taken only from software interfaces (SI), a client program installed on an occupant's PC. The SI detects if PC is used or not and that information is sent to the upper layer components as a sensor value (similar as for presence detection sensor). The sensor values are sent to the Gateway which unifies data and forwards to the Context component. The context component translates raw sensor data to message format understandable for the Decision Making component. The Decision Making identifies actions and sends them to the Orchestrator which ensures that actual environment adjustments are executed this time through the Sleep Management component.

Chapter 3

Implementing the GreenMind Smart Energy System

D esign of a system defines what are the components of the system, how are they connected and how they communicate. Implementation of a system determines which technologies are utilized and how the system behaves.

We present a prototype implementation of the GreenMind Smart Energy System[1], describing in detail the software implementation of its components and the implementation of overall system.

The choice of appropriate technologies is crucial for coverage of system requirements, especially non-functional ones, and that is the main focus of this chapter. Besides presenting and motivating the chosen technologies, we also presents technologies which serve as glue for the components, providing communication services. Furthermore, we present data model that enables unified and secure communication between components. Finally, as additional considerations, we present possibilities for extension of the system with other building sustainability-related applications, such as water and waste management applications, as well as applications covering the maintenance of the installed physical components in the field (i.e., buildings).

3.1 Implementation Methodology

The system is implemented using the principles of Service-Oriented Architecture (SOA). Moreover, even though the software development processes of the proposed system is described sequentially in this thesis, it is actually done using the Agile software development methodology (*Agile Manifesto* 2015). Per SOA

---

[1]The design and implementation of the architecture represent joint work with Brian Setz, Ilche Georgievski and Tuan Anh Nguyen, as well as a number of bachelor and master students mentioned in the Acknowledgments and the Related Work. We have collaborated with Viktoriya Degeler and Ilche Georgievski for obtaining the knowledge regarding the AI scheduling and planning techniques, respectively. We cooperated with Brian Setz on implementation of the computer sleep mode solution and the general framework, with Ilche Georgievski on implementation of the reasoning layer and with Tuan Anh Nguyen mostly on implementation of the physical layer.

methodology, each system component may have different implementation technologies as long as it behaves according to specifications and contains standard (defined) interfaces. The Agile methodology, is using Scrum as a framework for developing and sustaining complex products[2]. The Scrum is defined as a framework within which people can address complex adaptive problems, while productively and creatively delivering products of the highest possible value. Using Scrum, the software is developed by a number of Scrum teams working on various components, mostly in 1-4 weeks iterations (sprints). Each team gets frequent feedback from system owner and other stakeholders in order to implement the system that satisfies defined requirements.

3.2   The GreenMind Prototype Implementation

We implement a prototype system and deploy it in the Bernoulliborg building. In the prototype, we implement all components mentioned in the architecture, presented in Figure 2.1. Some components are visible and exposed to end-users, while other components, that are crucial for functioning of the overall system, are not visible at all. Our implementation involves software running in the cloud, on the thin clients responsible for sensor data collection and control of actuators, as well as software running directly on the sensors and actuators. We utilize several different programming languages and tools for implementing the different aspects of our system.

We implemented a separate *Gateway* for each used technology that can handle a wireless network of sensors and a wireless network of electricity measuring and control plugs, together providing enough ability to monitor essential environment information. In addition, we implemented a workstation software interface as another type of sensor/actuator that can monitor activity of computers and adjust their sleep mode. The *Facility Management Back Office* and the *User mobile Application* are closely related and developed as a part of the same development effort. The *Consumption Measurement* component is responsible for collection of consumption data from measuring plugs. The *Consumption Display* and the *Sleep Management Dashboard* represent the main interfaces to the end-users. The *Consumption Display* presents overall electricity consumption of a building, while the *Sleep Management Dashboard* shows current status and activity of PCs, together with historical data about the same. Other components that are less visible represent are the components of two middle layers, Utility and Reasoning layer. Complete implementation of *Context* component, *Database(s)*, *Decision making* component, *Orchestrator* and *Sleep Management* ensures that sensor in-

---

[2]http://www.scrumguides.org/scrum-guide.html

formation is properly gathered, processed, presented and finally controlled by actuators. The overview of the implemented system can be in the Figure 3.1.

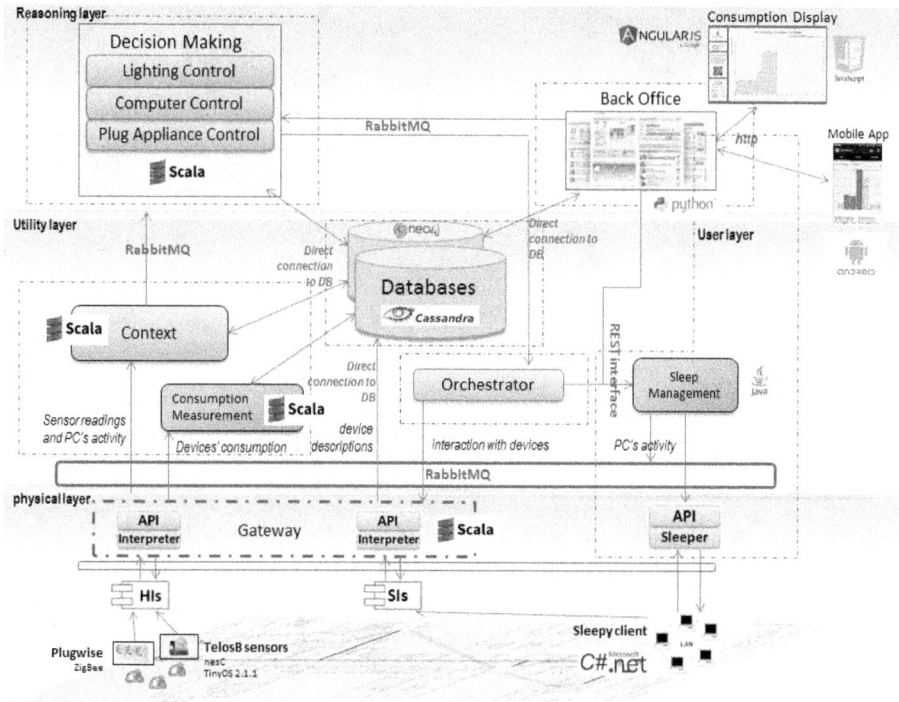

Figure 3.1: The overview of the implementation the GreenMind system

3.2.1 Physical Layer

The Physical layer is formed of *TelosB-based sensors*[3], *Plugwise sensors/actuators*[4], and the *Gateways*. These components together represent the connection of our system to an actual (physical) environment. So to say, they represent the eyes and the ears of the system. Using the components from the physical layer, we are able to understand ambiance conditions (level of light, movement of people) in any particular space in a building, without using any highly intrusive equipment (such as cameras).

In the current implementation, we use *IEEE 802.15.4* compliant wireless *TelosB*-based sensors produced by Advantic Systems (*Advantic Sys.* 2013), see

---

[3]https://telosbsensors.wordpress.com/
[4]https://www.plugwise.com/

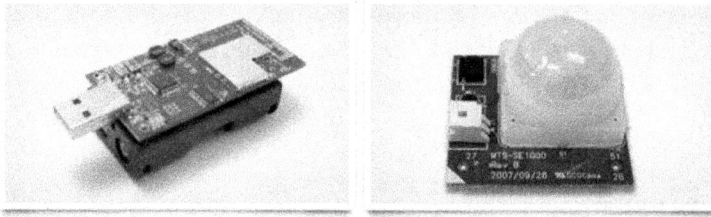

Figure 3.2: Temperature and Light sensor (left), passive infrared sensor (right)

Figure 3.3: Electricity measuring plugs

Figure 3.2. The on-board Passive Infrared Sensor (PIR) based motion detectors are used together with the light sensors. The motes, sensor nodes capable of processing, data gathering and communicating with other nodes in the network, are programmed in *nesC* and run on the *TinyOS 2.1.1* embedded operating systems. For electricity measuring plugs, we use *Plugwise* (*Plugwise* 2013) products consisting of plug-in adapters that fit between a device and the power socket, see Figure 3.3. The Plugwise adapters can "switch on" or "switch off" a plugged device, and they can, at the same time, measure the power consumption of the attached device. The plugs form a wireless ZigBee (*Zigbee* 2015) mesh network around a coordinator. The network communicates with the base station through a link provided by a receiver device (in a form of USB stick).

The technology choices are made based on previous experience with the above-mentioned sensors and actuators, time invested in development of drivers for the same and availability of devices. In theory, we can use any other sensors and actuators as long as the used protocols and message formats are known. At the end, the goal of our system is to be hardware agnostic, so we can utilize benefits from latest development of sensing and actuating hardware devices.

The software interface *Sleepy client* (in short, Sleepy) is written in *C#.NET*[5]. C# is an object-oriented language that enables developers to build applications that run on the .NET Framework. It can be used to create Windows client appli-

---

[5]https://msdn.microsoft.com/en-us/library/z1zx9t92.aspx

cations, XML Web services, distributed components, client-server applications, database applications, and other.

The Sleepy client runs on physical workstations (i.e., PCs) and is responsible for monitoring their activities. Currently, an activity is defined as keyboard and mouse movement. The Sleepy client is also responsible for adjusting the sleep timeout, a time period before a workstation enters sleep mode after being idle/i-nactive for a specified amount of time). The Sleepy client can also execute "go to sleep" and "turn off" commands if that is requested from the Sleep Management server.

Lastly, the Gateways are implemented as a process running in the background that reports state of controlled devices. The gateways are written in *Scala*[6] programming language. Scala is a programming language which runs on the JVM, that incorporates both object oriented and functional programming. Scala particularly excels when it comes to scalable server software that makes use of concurrent and synchronous processing, parallel utilization of multiple cores, and distributed processing in the cloud. It is used by many large companies, including Twitter, LinkedIn, and Intel.

3.2.2 Utility Layer

As previously stated, the Utility Layer is responsible for facilitating the main functionalities of the system, from environment sensing, through data storage and processing, to environment actuation. We present the implementation of five components comprising this layer.

Consumption Measurement

The main responsibility of the Consumption Measurement (CM) component is to gather electricity consumption data from the gateway and store it in a database. This component is internally also known as Plugwise Gateway. It was originally written in *Java*, and later on rewritten in *Scala*.

The CM uses the history buffer to limit the network traffic and processing load. In case of CM failure, the sensors (in this case Plugwise devices) continue to monitor and locally store the energy consumption data, and when the CM restarts it simply gathers all the missing data. The same applies when the network of the Plugwise devices is temporarily unreachable. Then, when the network connection is re-established the consumption component will retrieve all data until the last stored history log.

---

[6]http://www.scala-lang.org

Context

The Context component acquires, as its inputs, raw data from sensors and applies reasoning based on a set of pre-defined conditions over the raw data in order to provide high-level information about the environment that is used by the Decision Making component. For example, if the value of a light sensor equals 1000 lux, it is understood that the light condition is *LightLevel3*, meaning that there is enough natural light in the environment. Another example is fusing the raw data from a PIR sensor and a acoustic sensor in order to detect the presence of a user. It is envisioned that the Context component takes into account not only the data coming from sensors deployed in an environment but also feedback and preferences from users and learn over time, thus more meaningful information about users and a physical environment is acquired, providing better inputs for better environment adjustment strategies.

The ontologies (definition of the types, properties, and interrelationships of the entities) are modelled using *Protégé* (*Protégé* 2013), a graphical tool for ontology development that simplifies design and testing. Ontological reasoning is performed using the *HermiT* (*HermiT Reasoner* 2013) inference engine, and its application programming interfaces (APIs) for the Java programming language. The recognition algorithm is developed in Java and implemented as an on-line recognition system. More details are presented in (Nguyen, Raspitzu and Aiello 2013).

Databases

In our system, we use two databases, Neo4j and Cassandra (*The Neo4j database* 2014, *The Cassandra database* 2014). Neo4j is an open source database with features from both document and graph database systems. Neo4j excels in scalability, availability, performance, and price. After careful consideration, Neo4j is picked as the main storage facilitator for the building spatial and structural information of the GreenMind system, for example static information about the building, building organization (e.g., floors, rooms), sensors and actuators and their location, device and data type, etc.

As proposed in (Degeler 2014), to ensure future scalability of the system, we decided to utilize the benefits of a distributed fault tolerant database, Cassandra (Lakshman and Malik 2010). In our system, Cassandra is the main database, responsible for storing large amount of sensor data. Cassandra is a NoSQL database that delivers fast performance for systems involving time series or big data. It is used to store time series data coming from sensor readings periodically (e.g., each second one data point).

The Cassandra database[7] is optimized for scalability (*non-functional requirement NFR9, NFR10*) and high availability without compromising performance (*NFR6*). It has linear scalability and proven fault-tolerance on commodity hardware or cloud infrastructure. Cassandra is in use at over 1500 companies that have large and active data sets, including Apple (with over 75,000 nodes storing over 10 PB of data), Netflix (2,500 nodes, 420 TB, over 1 trillion requests per day) and eBay (over 100 nodes, 250 TB). This technological choice also supports non-functional requirement *NFR5* - fault tolerance. Using Cassandra cluster, data is automatically replicated to multiple nodes for fault-tolerance. Additionally, replication across multiple data centers is supported. Failed nodes can be replaced without downtime.

All data logging coming from different system components goes to the main database. This enables recording exact behaviour of a building, in other words building heartbeat, that is very helpful for further analysis of what causes energy consumption, for consumption prediction, and how the consumption can be kept in reasonable limits or even reduced using automated control.

Sleep Management

The Sleep Management component is responsible for collecting data from Sleepy clients and publishing this data on the messaging queue, RabbitMQ. It is implemented as a server component. Furthermore, it is also responsible for forwarding commands to the correct Sleepy client.

The Sleep Management component is written in *Java*. Currently, the Sleep Management fully supports Windows 7 and Linux (Xubuntu 14.04 LTS and Ubuntu 10.04 or higher), that are most widely used at the University of Groningen and also supported by its IT service department. For more details about the implementation of the Computer Sleep Management we refer to (Setz and Pul 2013) and (Hoeksema and Medema 2015).

Orchestrator

The Orchestrator populates a specific environment by retrieving the information from the databases. The set of variables, their types, locations, and properties are gathered from the static database (Neo4j). The initial values of variables are gathered from the dynamic database (Cassandra). Both databases provide a unified set of operations. Then, the orchestrator subscribes to the messaging queue, and awaits for messages, that is, events. The Orchestrator creates a domain object with the help of Planning services (Georgievski et al. 2013), and uses

---

[7]http://cassandra.apache.org

the problem converter to transform environment in a planning state. Upon each event, the Orchestrator creates an planning problem and invokes the core planning service. When a plan is found, if it exists, the Orchestrator translates the plan steps into acting services and uses the Gateway implemented as REST resources for execution. The Orchestrator is implemented in *Scala*.

3.2.3 Reasoning Layer

The Reasoning Layer represents the brain of the system. We present the implementation of Decision Making component and refer to work where more detailed descriptions are given.

Decision Making

The Decision Making component receives context information, it does its reasoning and provides a set of energy saving actions to be executed. This component decides what actuators should do on the physical level.

The Decision Making consists of the following components: *Problem Converter*, *Domain Modeler* and *Planner*. All components of the *Decision Making* (i.e., planning) system are implemented in the *Scala* programming language. The *Problem Converter* translates the context information and activities described in JSON syntax into Hierarchical Planning Definition Language (HPDL) syntax (Georgievski 2013). The *Domain Modeler* enables creating models. For the illustration, in Figure 3.4 we present an excerpt of domain representation from our graph database (i.e., Neo4j).

Consequently, the *planner*'s input consists of HPDL domain and problem descriptions. Planning is offered as a service by implementing its functionalities as REST resources. Upon receiving a request with appropriate arguments, the Planner searches for a solution, structures the resulting plan in JSON, XML or plain-text format, and sends it to the interested party, in our case the Orchestrator. More details about this component and how it utilizes AI planning HTN technique can be found in (Georgievski and Aiello 2015, Georgievski and Lazovik 2014).

3.2.4 User Layer

This layer contains interfaces for end-users and facility managers. We present the implementation of the Consumption Display, Facility Management Back Office, User Mobile Application and Sleep Management Dashboard. Besides implementation details we present the actual developed GUIs.

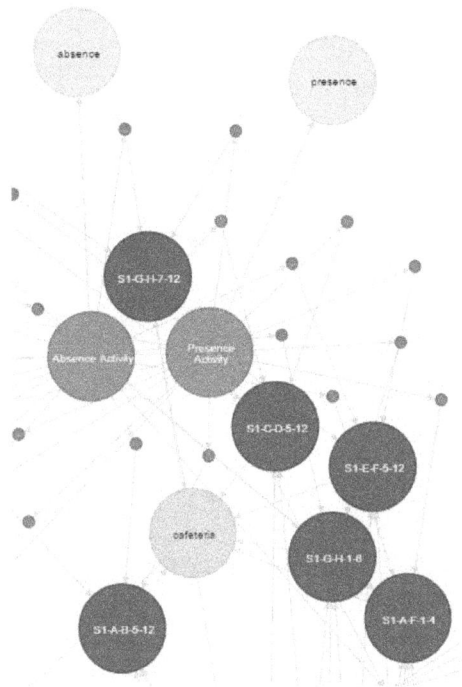

Figure 3.4: An excerpt of a domain model representation from the graph database

Consumption Display

The Consumption Display serves to present electricity (gas, water) consumption data to the end-users. The consumption display is developed using *AngularJS* framework[8], *Bootstrap 3*, and *HighCharts* graphs[9]. AngularJS is based on JavaScript programming language. AngularJS is client-side and and it is compatible with both desktop and mobile browsers. The Bootstrap 3 is used for one of the components of the consumption display. To show the real consumption data on the dashboard graphs, REST API is invoked. Every 5 minutes, a new call is initiated by an automatic refresh. To design the graphical interface and the incentives for the end-users, we consulted our colleagues from the Environmental Psychology Group of the University of Groningen, as well as professional designers. In Figures 3.5 and 3.6 we show design and implementation of the Consumption Display GUI evolved. To increase simplicity of the dashboard, described in non-functional requirement *NFR1*, we consulted the Faculty Commu-

---

[8]https://angularjs.org
[9]http://www.highcharts.com

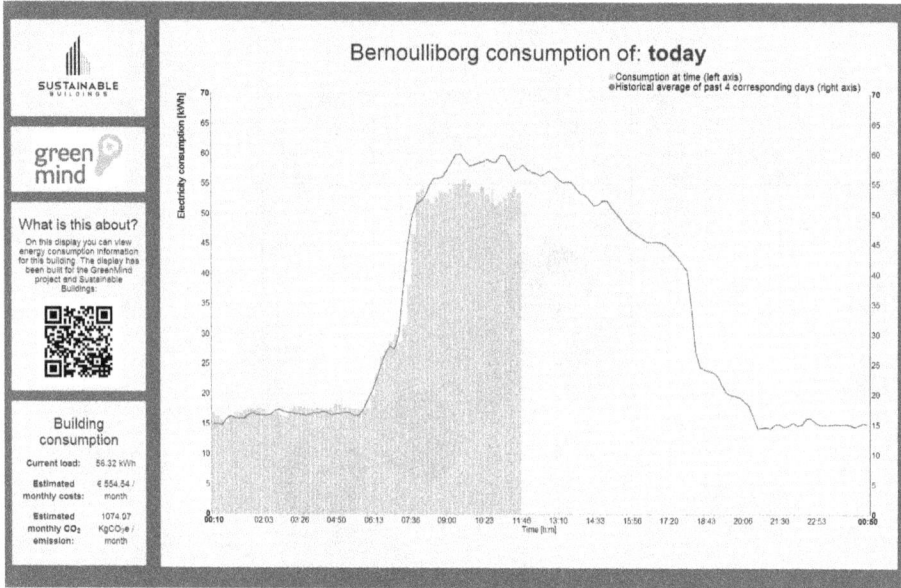

Figure 3.5: The Consumption Display 1.0 GUI

nication Department for feedback to improve the GUI and make it more simple
and understandable by the end-users.

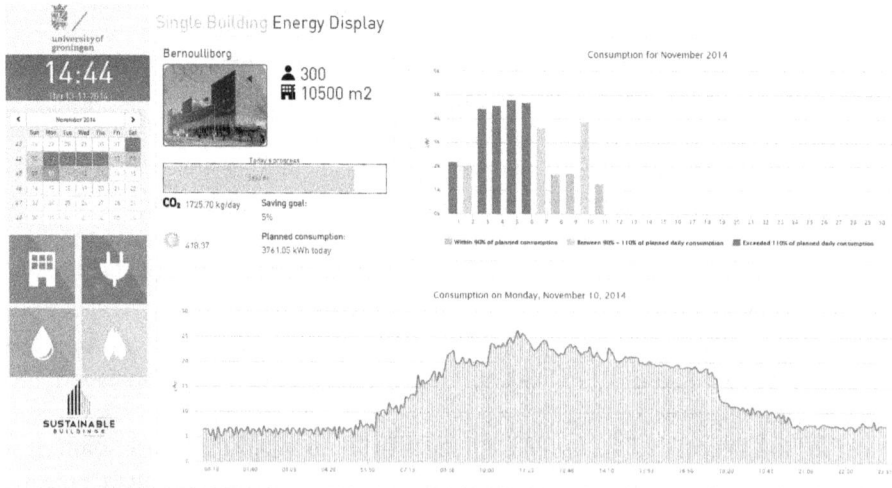

Figure 3.6: The Consumption Display 2.0 GUI (Energy Dashboard)

Facility Management Back Office

The Facility Management Back Office (in short, Back Office) is used for management of all data related to a building (building-floor-room), devices, users and groups. The Back Office has the following responsibilities: providing an Administration panel for configuration and viewing usage logs/statistic, aggregating and (pre)processing sensor data, providing Mobile Clients with an interface for authentication, authorization and data retrieval, caching of mobile clients requests, configuration of mobile clients (e.g. user account control) and pushing notifications (e.g. progress updates, achievements) to mobile clients.

The Back Office can be accessed by other components through a REST API using HTTP(S). Message content is represented using Application/JSON. Authentication and authorization are handled by passing a session token with requests.

The Back Office has been implemented using *Flask*. Flask is a microframework for Python based on Werkzeug and Jinja 2. A number of Python packages are used by the back office. The most notable of these packages are: *requests* (an HTTP client for Python), *pytest* (testing framework, with support for JUnit output), *kazoo* (Zookeeper client) and *flask-sqlalchemy* (ORM layer for database abstraction). More infromation about the implemetation can be found in (Meiboom 2013).

User Mobile Application

The responsibility of User Mobile Application (Figure 3.7) is to provide users with personalized usage statistics, general building energy use statistics, as well as notify users of progress on energy goals. It is believed that having this type of information on personal devices of the users could increase their interaction and motivation to contribute to energy conservation goals (Abrahamse et al. 2005, 2007). The User Mobile Application retrieves data from the Back Office through a REST API using HTTP(S). Message content is represented using application/json. Notifications are received from *Google Cloud Messaging*. The User Mobile Application depends on the Back Office for retrieving data as well as on the Google Cloud Messaging to receive push notifications. The current version of the application is developed for *Android* mobile devices.

Sleep Management Dashboard

The Sleep Management Dashboard (Figure 3.8) serves to identify all workstations that are being monitored and whose sleep timeouts are being adjusted by our Sleep Management server component. The Dashboard also shows status and

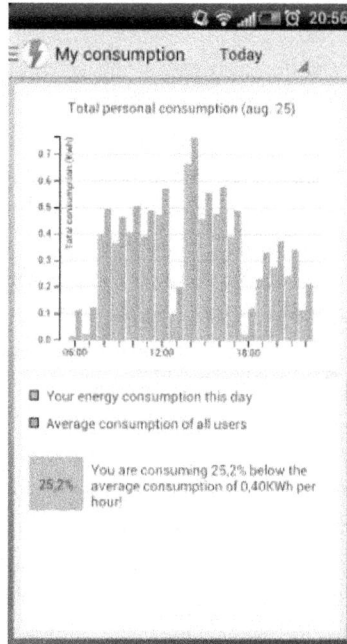

Figure 3.7: User Mobile Application

user activity on monitored PCs. The Sleep Management console is developed as
a web application using HTML, CSS, JavaScript, and Java.

Figure 3.8: Sleep Management Dashboard v1.0

3.2.5 System Communication

All the components in the GreenMind prototype system communicate with each other through the use of JSON objects. JSON is chosen because of the good readability, simple syntax and ease of use. JSON is also less verbose than alternatives like XML. Developers of components that communicate with each other dictate the content of the messages that are communicated with each other. Depending on the data that the component requires, the body of the JSON object may change. For the sake of traceability with asynchronous communication and readability and consistency in generic messages, a basic template is specified to which all the messages of components should adhere.

As stressed in (Degeler 2014), the amount of sensor data grows with the number of devices, and at some point, concurrency, queue processing speed and bandwidth issues may stop the system from functioning. Thus, to address this scalability issue on on the level of high-volume fast data processing, we choose to use RabbitMQ (Videla and Williams 2012). RabbitMQ messaging framework is chosen as it is a complete and highly reliable enterprise messaging system based on the emerging AMQP (*AMQP Messaging System* 2013) standard and runs on all major operating systems. It provides reliable ways for sending and processing large streams of data.

3.2.6 Data Model

To have unified communication, we also defined a general data model that enables inclusion of all different types of sensors. This means that we bring all sensor data in the same format that also allows us to reuse information among subcomponents, and that way better understand what happens in a building environment.

We use the following JSON format when publishing sensor data to the messaging queue. The structure of our data model is the following.

```
{"sensor_id":<UUID>,
"instance_id":<UUID>,
"timestamp":<timestamp>,
"value":<base64 encoded byte array>,
"process_id":<UUID>}
```

Let us explain each element of the defined data structure.

- *sensor id* represents an UUID identifying the type of sensor

- *instance id* is an UUID identifying the specific instance of the sensor to which the data belongs

- *logged at* is a UNIX timestamp of when the data was logged

- *value* is a Base64 encoded string representation of the binary array value being stored

- *process id* represents an UUID used to track sensor data generated from the same event (e.g., a user enters a room and this causes 3 sensors to generate sensor data, all these entries would share the same process ID). In the current implementation, even though accounted for, the *process id* is not yet utilized.

An example of transferred sensor data is the following.

```
{"sensor_id":"45dd228d-0627-4412-80e7-00b66a2ea8c0",
"instance_id":"da7369a6-27c0-47d9-96c6-cb3ecdc0da01",
"timestamp":1389470353562,
"value":"eyB0ZXN0IH0=",
"process_id":"2b6731d2-abce-444b-bea9-a5d29567e122"}
```

As it is visible from the data model description and the example, we use UUIDs, UNIX timestamp and base64 encoded byte array, and therefore to understand data from the database, some data conversions and decoding would be needed first. Base64 encoding is used for representing byte arrays as strings so that they can be efficiently transferred using JSON.

To publish data to queue, we use the following routing key format.

```
sensordata.<building>.<floor>.
<room>.<area>.<sensorid>.<instanceid>
```

Examples of routing keys would be the following.

```
sensordata.bb.5.*.*.sensor-uuid.instance-uuid
sensordata.bb.4.478.back.sensor-uuid.instance-uuid
```

The first example defines a sensor that works in the Bernoulliborg (in routing key: bb) the 5th floor, room and area are unspecified (i.e., defined as stars), and the second example defines a sensor installed in the back of room 478 on the 4th floor in the Bernoulliborg building. Each type of a sensor in the system has its own Sensor UUID and Sensor Name, as shown in Table 3.1.

Sensor Name	Sensor UUID
Sleepy State Sensor	920f1a76-940e-11e3-bca1-425861b86ab6
Sleepy Activity Sensor	1b9691c1-7ec0-11e3-baa7-0800200c9a66
Sleepy Disable Sensor	2c2b8c10-c3c2-11e3-9c1a-0800200c9a66
Lamp Status	bd38bdb0-a2f0-11e3-a5e2-0800200c9a66
Appliance Status	aac0f820-be6e-11e3-b1b6-0800200c9a66
Plugwise Electricity Consumption	e2a7b7e0-a2f0-11e3-a5e2-0800200c9a66
PIR Sensor	f553d6d0-a2f0-11e3-a5e2-0800200c9a66
Presence Activity	52d94550-ceda-11e3-9c1a-0800200c9a66
Absence Activity	d5f80460-cfb8-11e3-9c1a-0800200c9a66
Light Intensity Sensor	037e1720-a2f1-11e3-a5e2-0800200c9a66
Light Scale Sensor	52d96c60-ceda-11e3-9c1a-0800200c9a66

Table 3.1: Example of sensor names and sensor UUIDs

3.2.7   System Integration

The main strength of the system comes from its integration. By integrating different components, the sensor data that was not logged before the system implementation (for example, data from motion sensors that control office lights in a form of a closed loop) is now being stored and then reused among components for more accurate detection of the environment state, as well as better control of energy consuming devices.

In the sense of sensor data reuse, the system integration is done on the level of the database. That is where all the sensor data is being stored. The sensor data is retrieved by each component that needs that data to fulfil its functionality. For example, the Decision Making component retrieves sensor data that describes environment current state in order to decide how to control the actuators connected to energy consuming devices. That way, for instance, the information whether a PC is being used is exploited as additional information besides data from motion detection sensor, to determine presence in an office more accurately, and that way reduce unwanted switching of lights.

3.3   Additional Considerations

If we consider that an ICT system is a mean to achieve the goal of making buildings more sustainable, then we have to look wider then solely optimizing usage of electricity. We need to strive to minimize energy consumption (electricity, gas), but also to minimize waste production and increase recycling levels, as well as reduce water consumption. Therefore, the follow up goal could be to increase reuse of existing resources, increase harvesting of available natural resources as well as increase production from alternative energy sources. In the following, we present our initial prototypes working to support the first defined goal (reduction of consumption).

3.3.1   Water

General information about water use in the building can be measured, stored and displayed to building occupants. Sharing this information with users could raise their awareness and potentially result in reduction of their water consumption.

Therefore, we also proposed an architecture for a water consumption measurement and displaying system that is compatible with the our proposed Green Mind architecture. Moreover, we developed a prototype of the system. In Figure 3.9, we show our initial implementation of a front-end for the water dashboard. The system is described in detail in a master thesis (Musters et al. 2014).

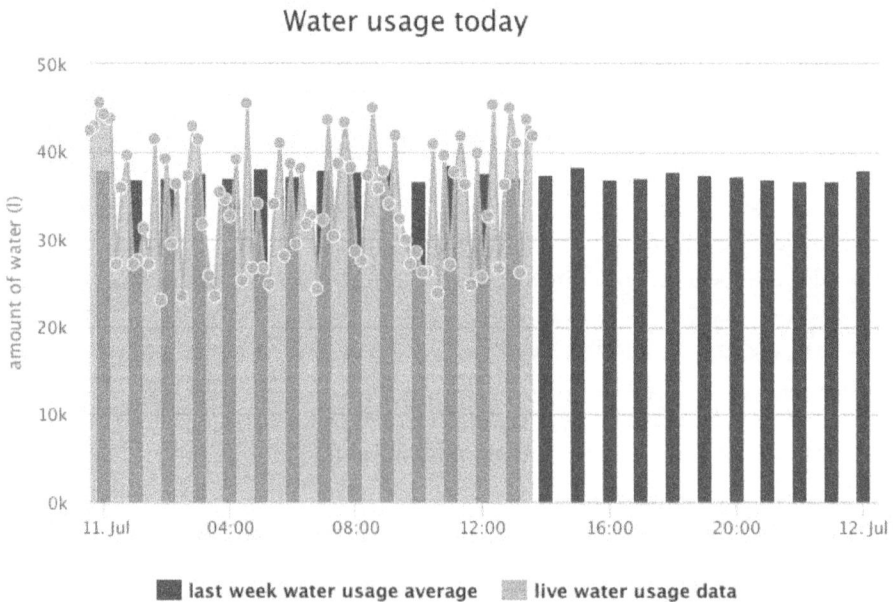

Figure 3.9: Sample graph, Water dashboard

3.3.2   Waste management

The waste management process can be included as a part of the building management system. That is why we proposed the *Social Recycling System* represents a recycling system that besides traditional incentives (such as money incentives) uses additional incentives, such as public recognition using the facilities of social networks, to increase users' extrinsic motivation for recycling. With this system, anonymous waste becomes personalized. That gives a solid ground for development of a system that makes recycling more interesting by incorporating competition aspects. Moreover, we propose hardware and software system that can support automation of this process.[10]

In Figure 3.10, we show how a sample graph from the recycling administrator dashboard. Detailed information about the implementation can be found in (Idsardi 2014). Moreover, in Figure 3.11, we present proposed hardware solution of a recycling kiosk.

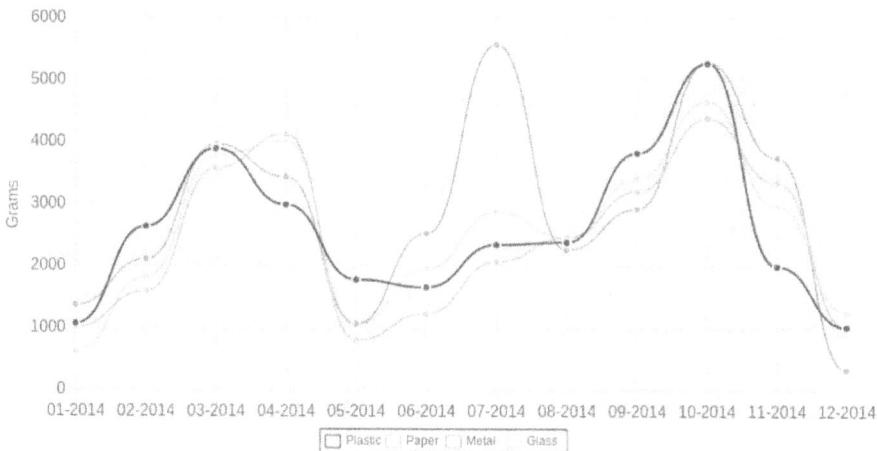

Figure 3.10: Administrator view, Recycling dashboard

3.3.3   Gas

We have shown how the information about electricity and water use can be measured and presented to users. One may draw the conclusion that the same solutions can be used for other purposes as well, for example, tracking and pre-

---

[10]Automation of waste separation and recycling process is presented in the manuscript "Social Recycling System" that is a joint work with Rešad Nizamić, who designed the proposed recycling kiosk.

Figure 3.11: Proposed hardware solution of a recycling kiosk. The solution is designed by: Rešad Nizamić.

senting gas consumption. Using the same infrastructure and implementation, we can acquire real-time information about gas consumption from gas meters, and by applying the same principles, present that data to the users, with the final goal of saving on gas within a building. Detailed information about how we can interface with gas meters and influence gas consumption is also described in our Green Mind Award 2014 winning project - Sensible Heating Operation Solutions[11].

3.3.4 Maintenance

Once all sensors and actuators, displays and other equipment forming a smart energy system are installed within a building, they have to be inventoried, periodically inspected and maintained. For that reason, we developed the Building Maintenance Application. With the Building Maintenance Application, all deployed inventory can be tagged and tracked. The location of each item is then known. This is especially important in case of failure of some of the equipment. The maintenance workers have to be able to easily to locate a piece of

---

[11]http://www.rug.nl/about-us/who-are-we/sustainability/green-mind-award/senseos.pdf

equipment having a defect. Also, some equipment may need to have periodic checks due to internal regulations or even legal ones. For that reason, we developed the Building Maintenance Application.

The Building Maintenance Application consists of a mobile and a server parts. The mobile part serves for in-field tracking of items as well as reporting of defected parts. The server part has responsibility of gathering and storing data regarding the inspected and maintained items. This enables functionalities such as generation of automated reports, and generation of periodic tasks for maintenance workers to perform. The mobile side of the Building Maintenance Application is implemented using the Android SDK. It uses MySQL lite DB for storing tasks locally. The server side of the application is implemented using Java, Spring Framework and Kundera. It does automatic conversion from Java objects to NEO4j nodes and Cassandra entries. Detailed information about the implementation can be found in (Jans 2015). In Figure 3.12, we show the front-end of the first version of the Building Maintenance Application mobile application.

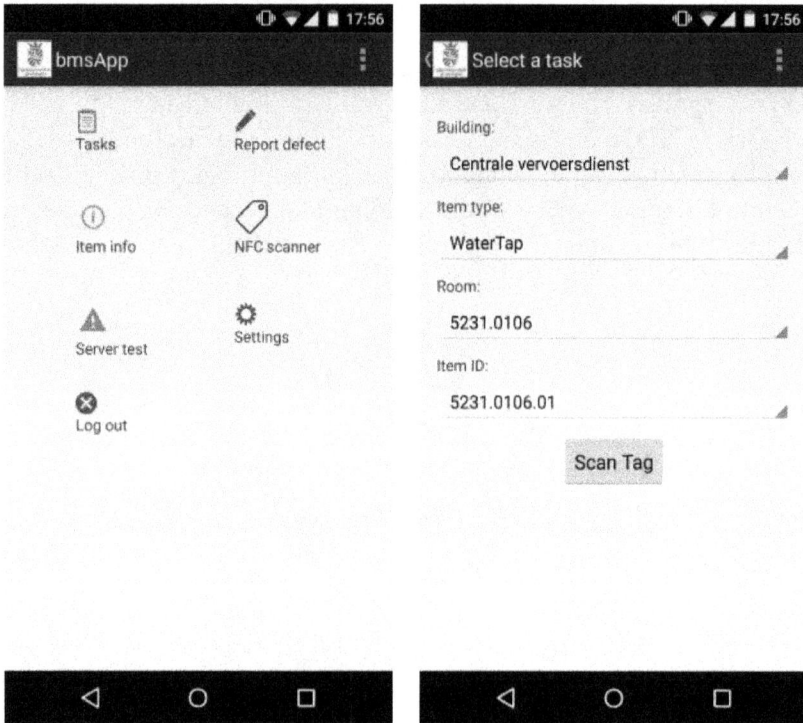

Figure 3.12: Mobile App, Building Maintenance Application

Chapter 4

Deployments of the GreenMind Smart Energy System

D eployment in operating environment presents many challenges, most of which are unexpected. We present how each component of the implemented architecture is integrated and deployed in the Bernoulliborg.

Deploying a software system in an operating environment, such as an University building, besides technical, brings organizational and communication difficulties, especially when it comes to software that interferes with an environment occupied and used by people. The challenges may be of different nature, from getting approvals for deployment, through legal regulations regarding equipment installation, all the way to mechanisms to protect people's safety and privacy. Moreover, cooperation with other departments responsible for installation and maintenance (such as Facility management or IT services department) has to be established. Equipment such as sensors, actuators, thin clients, servers, etc. has to be purchased and delivered by suppliers. This kind of dependencies on external parties may result in unwanted realization delays.

Additional work on implementation may be required as well. Once we encounter the limitations in an operating environment, we may notice that some of our previously developed solutions need additional adjustments to fit the existing infrastructure. For example, the display resolution of the energy dashboard presenting a building consumption may need to be adjusted to fit the previously installed or new displays; or software for the control of workstations' sleep mode need to be adjusted to be able to install it to both workstations managed by IT services department, and those that are directly managed by occupants themselves.

However, once we overcome the mentioned hurdles, necessary equipment can be placed in the operating environment and implemented software can be deployed on it. This leads to system integration. "In engineering, system integration is defined as the process of bringing together the component subsystems into one system and ensuring that the subsystems function together as a system" (Gilkey 1959). During system integration, all components are connected to each other in one platform. Usually, the integration process reveals additional adjustments that have to be implemented. The integration happens between the "neighboring" components, most often inter-dependent components. Therefore, inter-dependent components should also be *visible* to each other, in technical terms, they should be on the same network or sometimes even in the same sub-network. For example, in order PCs to be controlled by the system, at least Sleep Management and Orchestrator components should be deployed on the same sub-network as PCs, so they are able to exchange messages.

The deployment can be done in phases. As there are more different energy consuming subsystems (e.g., lights, workstations, appliances, heating, etc.), a team responsible for realization may decide to tackle one subsystem at a time. Therefore, only components that ensure full functioning of a subsystem *sense-decide-control* loop have to be installed first. For example, to control the workstations, we need to have deployed a client application on each controlled workstation, a server application that gathers data and decides how to control, a database where all activity and control data is stored, and a messaging queue application responsible of ensuring the availability of the required communication between the components.

When talking about deployment, one has to be aware that it includes also work at different locations of a building (offices, meeting rooms, restaurants, etc.). Therefore, occupants that inhabit these spaces also have to be timely informed and involved in deployment process. As we control different energy-consuming subsystems, we defined several different locations to be our living labs, where the subsystems work in environments occupied by users. It is important to mention that the smart energy system is deployed on top of the existing building automation system. In the following, we describe locations used as living labs, as well as how each *functional group* is deployed at those locations.

4.1   Living Labs

In total, we use six spaces as living labs. Three living labs are used during development and testing of prototypes, and another three are used at a later stage for final deployment of solutions in the operating environment. In Table 4.1, we list the living labs, their location, area, and number of devices deployed per living lab.

Three labs that are located on the 5th floor of the Bernoulliborg have in their name the prefix *DS*. The DS prefix stands for *Distributed Systems*, as the labs are created in space used by the Distributed Systems research group[1]. Therefore, we

---

[1]http://distributedsystems.nl

Living Lab	Location	Scale	Devices
DS Offices	5th floor	12 rooms	45 sensors
DS Lab	5th floor	1 room	7 PCs
DS Social Corner	5th floor	1 room	1 display
BB Entrance	Ground floor	1 large room	4 display
BB Workstation	Offices	14 offices	14 PCs
BB Restaurant	Ground floor	1 large room	45 sensors and actuators
Total number:		**30 rooms**	**116 devices**

Table 4.1: Living Labs in the Bernoulliborg building

first list DS living labs.

The DS Offices lab is spread across 12 different rooms on the 5th floor, including offices, a meeting room, a social corner and a kitchen. In this lab, we have 45 measuring sensors. The purpose of this lab is monitoring personal consumption of users, as well as monitoring joint consumption per device in the meeting room, social corner and kitchen.

The DS Lab is located in one room on the 5th floor. There, we have 7 controllable workstations installed. The purpose of this lab is monitoring and operating workstations consumption in a controlled environment.

The DS Social Corner lab is one room where kitchen equipment and lounge furniture are located. In this space, we placed a monitor to display common consumption to get end-user feedback on the energy dashboard during the development phase.

The rest of the living labs are located in the spaces used by the general building occupants and building visitors. *BB Entrance lab* is one large room located at the entrance of the Bernoulliborg. There, we used existing displays to show building common energy consumption in a real working environment.

The BB Workstation lab is located in the private offices throughout the building. Software interfaces controlling individual workstations are deployed in 14 offices, covering 14 workstations. The purpose of this lab is monitoring and operating workstation consumption in a actual working environment.

The BB Restaurant lab is located in a large room on the ground floor, and shown in Figure 4.1. There, we installed 45 sensors and actuators. 30 sensor-actuators are in charge of measuring consumption of the 30 light fixtures, while 15 sensors are responsible for detecting presence in different areas and measuring light levels.

Figure 4.1: Deployment of sensors in the Bernoulliborg restaurant

4.2   Users

The users occupying living labs differ. In the *DS living labs*, users are mostly researchers and a few staff members responsible for the administration (e.g., secretaries). In the *BB Workstation lab*, the users are both staff and research members who possess a workstation and who voluntarily applied to be part of energy saving experiments.

The *BB Entrance* and the *BB Restaurant* labs comprise the same connected space and therefore the same number and type of users applies (occupants and visitors of the Bernoulliborg). The approximate number and type of users per living lab is shown in the Table 4.2.

In total, we estimate that 886 users are affected by the system, mostly being students, researchers, faculty employees and visitors. Even though the capacity of the restaurant is only 200 persons at the same time, we estimated that about 800 users circulate during the lunch period.

4.3   Deployed Solutions

We have developed, tested, and deployed a number of energy saving solutions, including: (1) Consumption measurement, (2) Consumption display, (3) Computer Sleep Mode control, and (4) Lighting control. The techniques and al-

Living Lab	Aproximate no. of users	User type
DS Offices	15	Researchers, administrative staff
DS Lab	7	Researchers
DS Social Corner	50	Researchers, administrative staff
BB Workstation	14	Faculty employees
BB Entrance	*800	Students, employees and visitors
BB Restaurant	*800	Students, employees and visitors
Total number:	**886**	

Table 4.2: Approximate number and type of users per living lab

gorithms used to develop these solutions are described in Section 3.2.

In most cases, the deployment is done using *Docker*. Docker is an open platform for building, shipping and running distributed applications[2]. It gives programmers, development teams and operations engineers the common toolbox they need to take advantage of the distributed applications. Docker allows composing applications from micro-services, as well as the ability to deploy scalable services, securely and reliably, on a wide variety of platforms. It provides application portability, by packaging application, dependencies and configurations together to ensure that an application will work seamlessly in any environment on any infrastructure. Using Docker, we satisfied non-functional requirements *NFR2, NFR3, NFR4 and NFR11* (Section 2.2.2).

In this section, we give more details on the deployment of each solution. Then, we present the setup and concrete information regarding the deployed infrastructure.

4.3.1 Consumption Measurement

The Consumption Measurement system is responsible for collecting data about the real-time energy and other consumptions, and transfer the data to the database. This enables later presentation of the data for purposes of data analysis or provision of feedback to system maintainers and system end-users.

As a part of the Consumption Measurement system, we deployed two solutions: one for individual energy consumption and one for overall building consumption measurement. To understand individual energy consumption and its patterns, we deployed electricity measurement devices in 12 areas on the 5th floor of the Bernoulliborg. Layout of the consumption measurement lab is illus-

---
[2]https://www.docker.com

trated in Figure 4.2 and Figure 4.3. The areas that are measured by the system include eight offices, two hallways, one kitchen area, and one meeting area. Over the course of six weeks, from mid May to late July 2013, the office environment on the fifth floor of the Bernoulliborg building has been monitored 24 hours a day and 7 days a week. The information from Consumption Measurement was used to develop the BackOffice component which serves to model building spatial data and to assign particular measurement devices to users, users to rooms, etc.

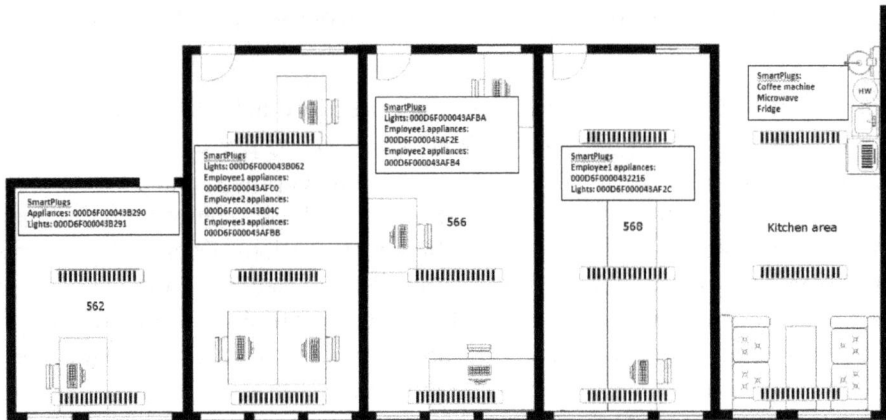

Figure 4.2: Consumption measurement - rooms 562, 564, 566, 568 and the kitchen area

The Consumption Measurement component is deployed in the operating environment and at the moment of writing this thesis it is working for more than one year. In the first deployment, it was deployed as experiment to determine potential savings per type of energy consuming device (Jilings and Heitmeijer 2013). Electricity consumption per measuring device (plug) was collected with frequency of 5 minutes and stored in the database. One of the conclusions of this work was that installing a building management system that controls multiple occupied offices has the potential to save energy. More specifically, the estimated energy savings from lighting range from 36% for corridor lighting to 56% for office lighting. Energy can be saved also on the appliances, namely, 56% for the coffee machine, 28% for the water boiler, 22% for a regular workstation, 9% for the fridge, and 7% for the microwave. When these potential energy savings are combined they reach average a potential energy saving of 27.5%. If the excesses of the research workstations, that are operated 24 hours a day and 7 days a week, are taken into account the potential energy savings are 15%.

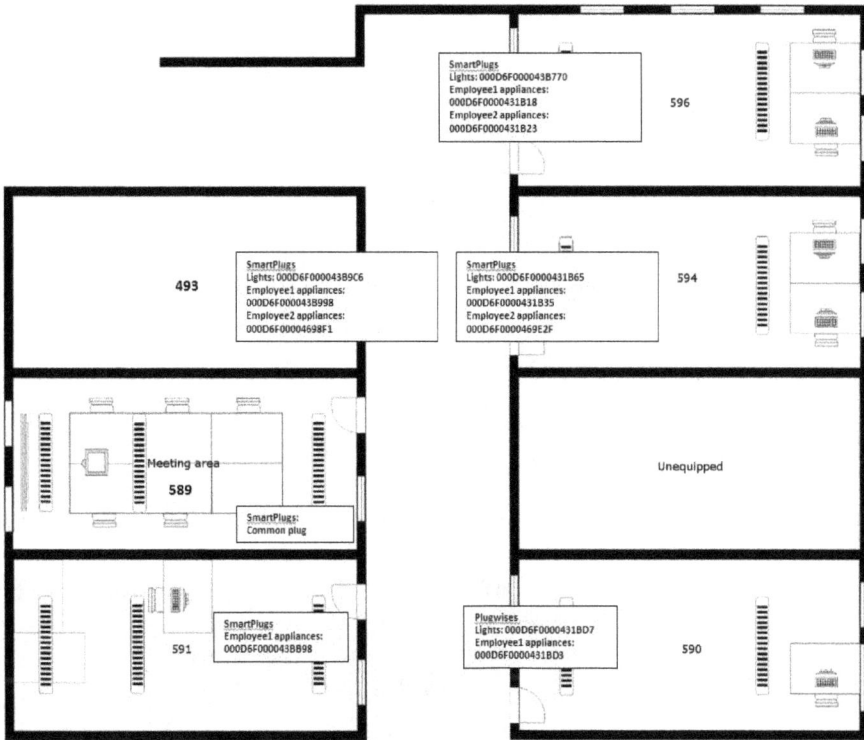

Figure 4.3: Consumption measurement - rooms 493, 590, 591, 594, 596 and the meeting area

Subsequently, consumption measurement component was revised and deployed to collect electricity consumption from employees with the purpose of displaying consumption data in a gamified way, with the purpose of motivating employees to perform energy saving actions (Dekker and Brandsma 2015). The Energy Competition Dashboard is illustrated in Figure 4.4.

Besides individual electricity consumption measurement, we also measured overall consumption for the whole building. To do that, we developed a hardware solution, *Smarter Meter* which is deployed to the building's main electricity meter, as shown in Figure 4.5. This solution ensures data collection from the meter's pulsing led lamp using optical sensor and its transfer to our database. Once the pulse data is in the database, it is converted to more meaningful units (i.e., kWh) and ready to be shown in presentation layer (e.g., energy dashboard).

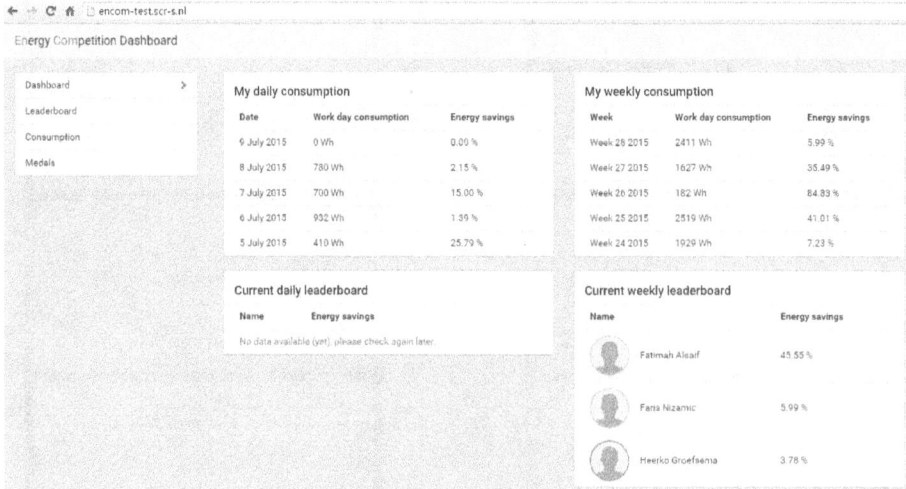

Figure 4.4: Energy Competition Dashboard for occupants of the Bernoulliborg

Figure 4.5: Smarter Meter deployment on the main electricity meter

4.3.2   Consumption Display

The Consumption Display has responsibility to present consumption data to system maintainers, end-users, and other interested parties in simple and usable way, that enables fast understanding of consumption, as well as to give feedback and ideas about eventual conservation actions.

To understand the patterns of electricity and other consumptions at the level of the building as a whole, as well as to present the building energy consumption to occupants, we developed a consumption display and deployed it in the Bernoulliborg building. At first, the consumption display was placed at the *DS Social Corner* lab, (Figure 4.6), and once proven that the solution is stable enough, it was deployed to BB Entrance and other screens on the first floor (Figure 4.7).

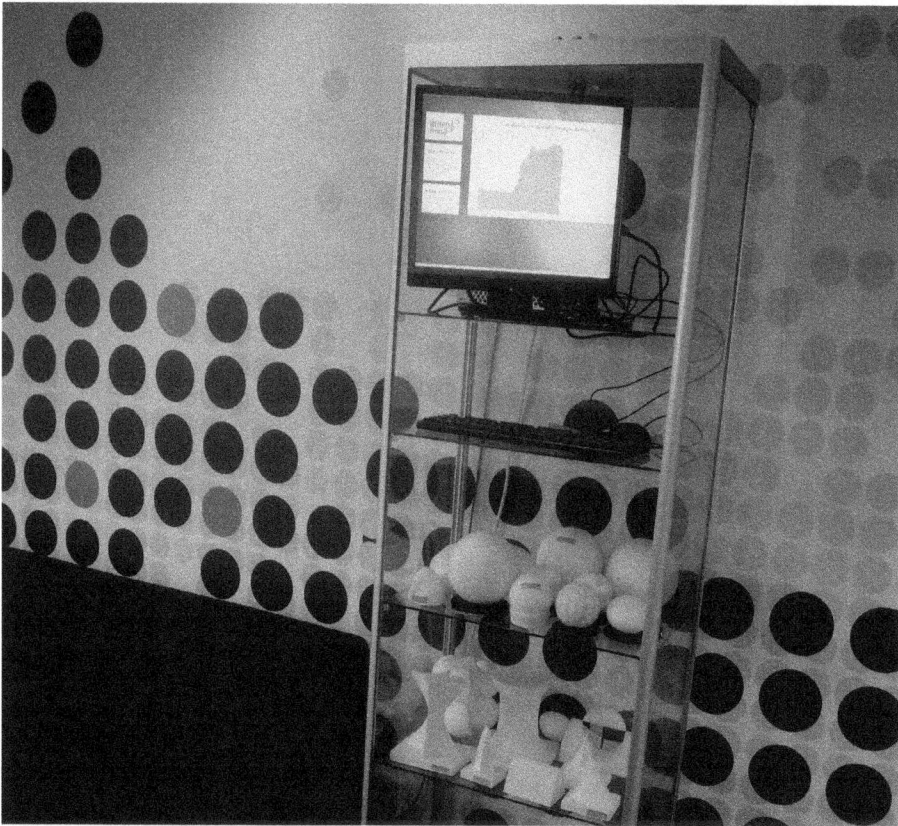

Figure 4.6: Test consumption Display at the 5th floor of the Benoulli building

The information shown on the consumption display consists of a daily view and a monthly view of the overall building electricity consumption, together with gray lines showing planned daily consumption. The planned consumption is calculated as defined percentage of expected saving (in our case 3 percent) of average consumption for that particular day. The average is calculated as an average of all the same days for as many years stored in historical data (in our

Figure 4.7: Consumption Display at the ground floor of the Benoulli building

case 5 years). The calculated value is marked as the planned consumption and taken as a daily goal to be achieved.

We use this mechanism to raise awareness on energy consumption, as well as to try to change the behavior of of building occupants. This is achieved by providing them with an energy saving goal and suggesting concrete saving actions that they can do in order to contribute to this goal. If the actual daily consumption is below 90% of the planned or aimed one, then we color it green, indicating that the goal is achieved. Between 90% and 110% it is colored yellow. Everything above 110% of planned daily consumption is colored red, indicating that the goal is not achieved. To serve the same purpose, that is, to motivate energy preservation behavior of occupants, we also display the amount of money that needs to be payed for consumption that day, amount of CO_2-equivalent emitted (in kg/day) as well as the saving tip of the day (e.g., "Please remember to turn off your pc when you have finished working").

The consumption display is developed in two variants, one for building occupants and one for facility managers. The *variant for building occupants* is optimized for large screens, without scrolling possibilities, containing energy saving incentives that may motivate building occupants for action. The *variant for facility managers* contains more advanced options, such as integration with weather information and occupants presence data, as well as access to control dashboards.

4.3.3 Lighting Control

The Lighting Control solution has responsibility to control lights based on environmental data collected by sensors (e.g., PIR and LUX sensors), with the goal of reducing energy consumption whenever and wherever that is possible. The Lighting control system is deployed in the restaurant on the ground floor of the Bernoulliborg. The restaurant is located on the ground floor of the Bernoulliborg. The restaurant covers a total area of 251,50 m^2 and has a capacity of 200 sitting places. The restaurant has glass walls from three sides, providing a significant amount of natural light when the weather conditions allow for it. The restaurant area is used for lunch from 11:30 a.m. 2:00 p.m. Outside these hours, the area is used by staff, students, or other visitors for working, meeting, or other social activities.

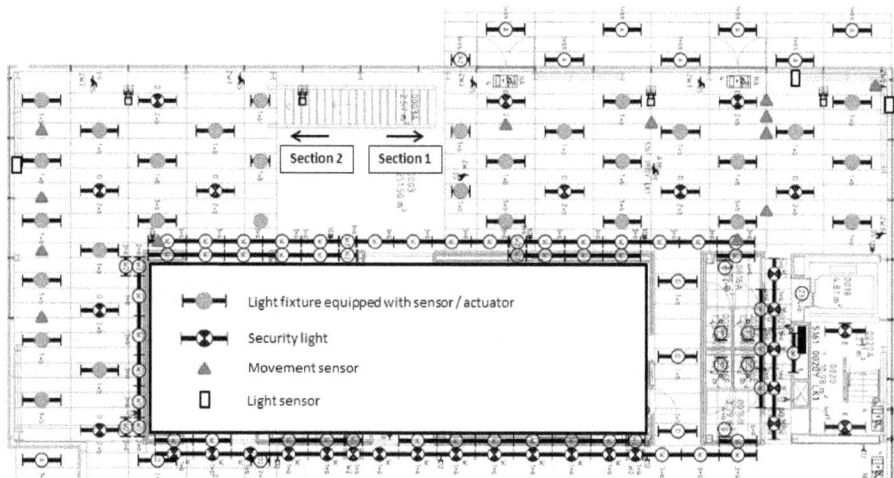

Figure 4.8: Placement schema of the sensors and actuators in the restaurant at the ground floor of the Bernoulliborg

The restaurant area is an open space divided in two sections. We make use of this division in the deployment. The layout is illustrated, together with the locations of deployed sensors and actuators, in Figure 4.8. In particular, each section has 15 controllable light fixtures, making 30 in total. There are several light fixtures that are uncontrollable and represent security lamps. While we do not control these, we take into account the light that they provide. In addition, there are two types of controllable fixtures. The first ones are large and have 38W of power consumption each, and small ones, each of which has 18W. These fixtures are controllable using the actuators attached to them, which also serve as sensors by providing information about the fixture's power consumption. We installed 15 sensors, three to measure the natural light level, and the rest to detect people's movement. In order to make more meaningful use of the restaurant space given the movement sensors, we divide each section into smaller spaces, called *areas*. In our case, a section comprises of more areas. In each area, we embed a movement sensor to understand environment conditions (e.g., level of light, and presence or absence of occupants within area).

We attached sensors and actuators to each individual lamp, marked as full circles in Figure 4.8. Using the sensors and actuators, we measure the consumption as well as control each individual lamp or group of lamps. Besides that, we also deployed motion and light sensors (in the figure represented as triangles and rectangles, respectively). These sensors give input to the Decision Making component which decides when the lights do not necessarily need to be on (e.g., when no one is present in a particular area, or when natural light provides satisfying level of light). Using this system, we reduce electricity consumption by using the lights only when they are absolutely necessary. For the restaurant area alone, the experimental results show that electricity savings of 86% were achieved.

4.3.4   Computer Sleep Mode Control

The Computer Sleep Mode Control has the responsibility to gather data about the status of computers and their usage, with the goal of adjusting sleep mode timeout value to the minimum user-satisfying value to introduce additional energy savings from computers. In order to understand the patterns of workstation usage, sleep timeout preferences of employees and part of the consumption coming from workstations, we have deployed Computer Sleep solution initially for 7 workstations at the DS Lab, and later on to the BB Workstation lab covering 14 workstations within the 14 offices in the building.

Per monitored workstation, we show the activity (working or idle) and status (off, on, sleep). Besides that, we automatically adjust the sleep timeout of workstations (*non-functional requirement NFR7*). We developed features that enable remote execution of commands such as: 1) send to sleep, 2) wake up, 3) turn off, and 4) turn on, but they are disabled as we want to minimize interference with employees work on their PCs. In other words, we only adjust the sleep timeout value and leave the operating system to send a PC to sleep mode once the operating system concluded that there is no active (e.g., using keyboard or mouse) nor passive interaction (e.g., watching movie, listening to music, etc.). The activity and status of workstations are shown on a dashboard, Figures 4.9 and 4.10.

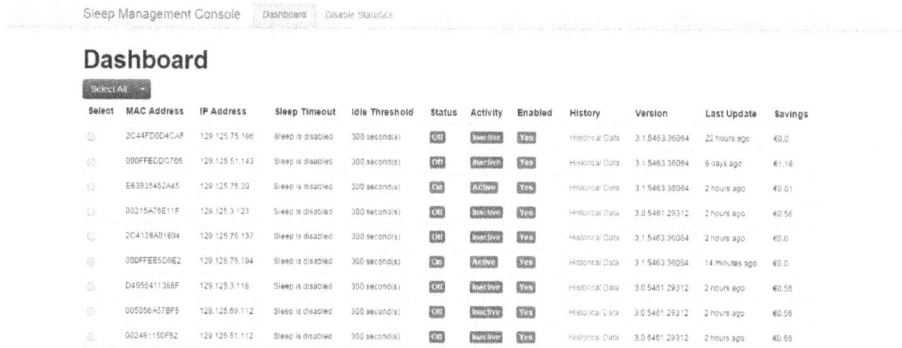

Figure 4.9: Computer Sleep Mode - Control and Analysis Dashboard

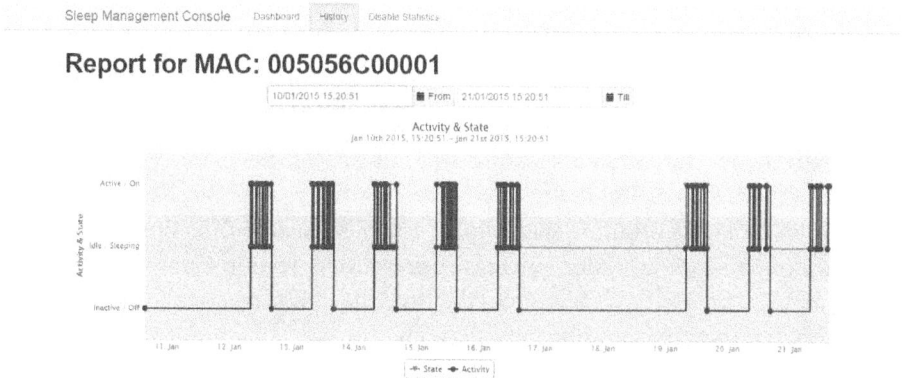

Figure 4.10: Computer Sleep Mode - Activity and State Analysis Dashboard

Historical data is also kept in the database. Historical data serves to determine user consumption from workstation as well as electricity savings using our automated sleep timeout adjustment. For the algorithms we use, the minimum timespan for data collection which is needed for learning the user behavior is one month. From the database information, we generate various reports which serve to inform facility managers and to motivate building occupants about further energy preservation.

4.4   Additional Considerations

During the realization of the project to make the Bernoulliborg more sustainable, several retrofitting solutions were proposed and realized. This work led to efficiency improvements, energy and water savings, and waste reduction. These solutions are realized by a multidisciplinary team including project leaders, computer scientists, software developers, university sustainability and energy managers, building and facility managers, demand manager, behavioral scientists, communication officer, installation and maintenance workers, electricians, etc.

4.4.1   Electricity Consumption Reduction using Sensor Holders

The Sensor Holder solution (Figure 4.11) enables adjustment of sensor direction with the goal of increasing presence detection and reduction of a sensor timeout. This leads to energy use optimizations by providing lights only when they are needed for the period they are needed. Using this solution we reduced the consumption of office lights in the Bernoulliborg. Previously, most of the movement sensors were located far from the location where occupants were performing their work. That way, sensor timeouts were too long, causing the lights to stay on unnecessarily for 30-45 minutes after the occupant had left the office.

We understood that fixed sensor positioning in offices is indeed an issue as their visibility range is limited and location of workers may be changed in time. That is why it is important to have ability to adjust sensor direction, to increase accuracy of presence detection. Increased presence detection gives ability of sensor timeout reduction, which reduces the time the lights stay on when that is not necessary (i.e., where is no-one present in an office).

Now, lighting within the Bernoulliborg offices is optimized by developing a hardware solution, an intermediate sensor holder, and deploying it to existing movement sensors in each office of the building, in total 180 offices. This sensor holder makes it possible to point the movement sensor in the direction of an employee, which gives better presence detection as well as the opportunity to reduce the light timeout to a much lower value. The turn off light timeout has been reduced from 45 to 15 minutes, which leads to monthly savings of about 2 hours for lighting per office per day.

4.4.2 Water Consumption Reduction using Flow Reducers

The purpose of the Water Consumption Reduction solution was to collect data on water consumption, to provide understanding how water is consumed within a building as well how this consumption can be affected, both by retrofitting as well as providing feedback to end-users, mainly responsible for this consumption.

Water saving devices (Figure 4.12) have been installed on all the water faucets within the building. Previously, a lot of water was unnecessary flowing when the building occupants washed their hands or their dishes after lunch. Thirty-five water saving aerators in total have been added to the faucets and this way the water flow from faucets is reduced by up to 50%, that later measurements showed represents 5% of total water consumption in the building.

4.4.3 Waste Separation Process Change

The purpose of the Waste Separation Process Change was to collect data on separated waste, and to observe possible optimizations from the change in this process.

In the Bernoulliborg, the waste separation process has been completely changed. Instead being able to dispose only general waste and paper, the occupants of the Bernoulliborg now have the opportunity to separate two additional types of waste (plastic and cans). Thirty new waste separation bins (Figure 4.13) have been deployed across the building and all old bins for general waste have been removed from the offices and common spaces.

Figure 4.11: Deployed intermediate sensor holder

Figure 4.12: Water faucets with water flow reducers

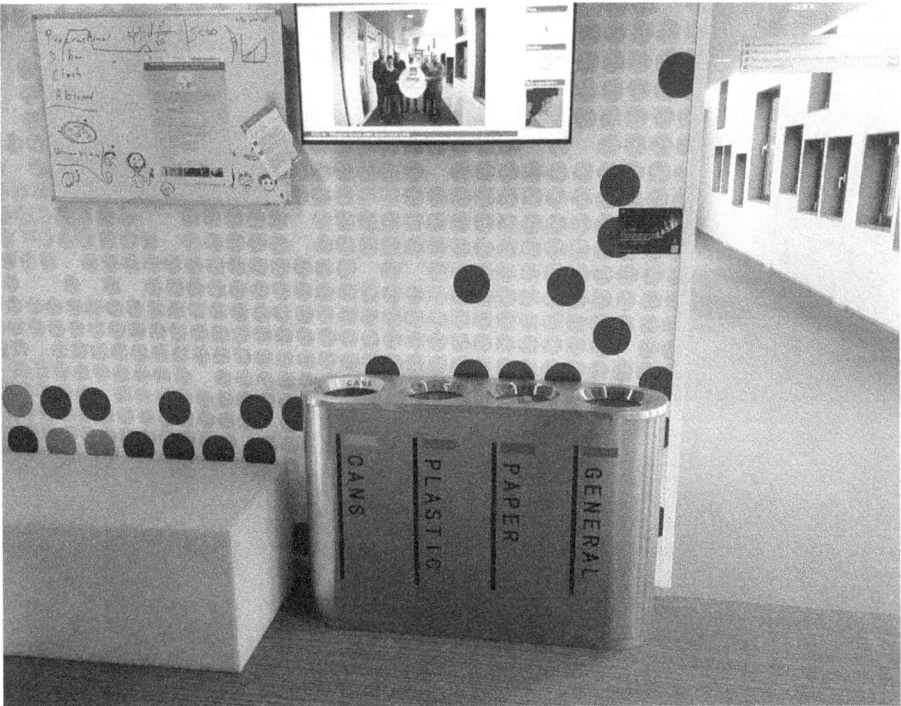

Figure 4.13: New waste separation bins

Chapter 5

Optimizing the GreenMind System Infrastructure

The proposed GreenMind system is used to optimize the use of resources in buildings. However, it is also important to optimize resources used by the system itself. This is relevant for a number of reasons. Firstly, to make the system economically sustainable, we have to support requirements coming from the business side, such as *cost effectiveness* and *scalability*. Moreover, the system should also support sustainability principles itself. In other words, an organization providing such a system should use the minimum resources required for its proper functioning.

To support these business and sustainability requirements (*NFR9 and NFR10 - Scalability, BR1 - Cost effectiveness, BR2 - Controllability*), we decided to use cloud-based infrastructure as one of our technological choices. Clouds are defined as a large pool of easily accessible virtualized resources (such as hardware, development platforms and/or services) which can then be dynamically reconfigured to adjust to a variable load, allowing for optimum resource utilization (Vaquero et al. 2008).

Cloud services are used for storage and data processing. The GreenMind system collects, stores and processes a large amount of sensor data, and that is an additional reason to use cloud services. Moreover, the cloud has high availability from any location, which enables central data storage and large data processing. More importantly, as the cloud has an essentially *unlimited* amount of resources, it can easily support the technological and business scalability. Clouds also scale cost efficiently. In other words, the use of cloud resources can be requested and its use terminated at any time, while the resource requester only has to pay for the time that the cloud resources were used.

Independently from the fact that the cloud is abstractly described as an unlimited source of computing resources, it is important to understand that in reality a cloud consists of actual physical resources (e.g., servers), which use energy, cost money and contribute to greenhouse gasses emissions (Pernici et al. 2012). For large companies that are using a huge number of cloud resources, energy consumption plays an important role. As an illustration of a possible magnitude of consumption, a Google data centre consumes as much power as a city the size of San Francisco (Buyya et al. 2009).

Large organizations such as universities and governments use hundreds of buildings for their operations. As we expect to serve such organizations, our system will need to support hundreds of buildings with different services. Increasing the number of buildings to be supported will cause significant load on the system and will require a lot of computing resources. Even though energy and economic savings generated by internal infrastructure optimizations may be significantly smaller compared to optimizations achieved in the field (i.e., in the clients' buildings), the optimization will lead to lower costs of infrastructure, higher economic acceptance and lower environmental impact.

To fully exploit the benefits of the cloud, one still needs to carefully manage the dependencies between different parts of the system to be deployed. This ensures that all services are in place, resources that are not required are not launched (or are shut down) to achieve near-optimal resource usage. The manual managing of an increasing number of services may be also very difficult and can lead to additional inefficiencies. Still, practice shows that many large companies that are working with sensitive data schedule cloud resources manually. Without automated scheduling, managing dependencies is error prone and very complex for humans, and potential savings are hard to be achieved. Furthermore, resource requests coming from different users may not be handled in a fair and optimal manner. A resource requester can be anyone requesting business services, e.g., a customer requesting different services, such as reporting or benchmarking for the purpose of an inspection. Frequently, consensus about the actual priorities for resource utilization may not be achieved, and the resources are allocated in a first-come-first-served or some other more arbitrary way. These arbitrary decisions lead to a non-optimal resource utilization and, therefore, put additional costs on a company.

To optimally deploy several kinds of client service requests, we use two AI techniques, *scheduling* (Degeler 2014, Nizamic et al. 2012) and *planning* (Georgievski 2013, Georgievski et al. 2013, Georgievski and Aiello 2012). Examples of those requests are *temporary use of energy dashboard*, a service of *performance benchmarking of buildings* within an organization, *periodic consumption prediction, fault detection in installations*, various *reports generation, execution of learning algorithms*, etc. These activities can be initiated from several groups, requesting different amounts of resources that have to be used for different amount of time. For example, a service of consumption prediction can be periodically requested before building performance meetings. Another example is a request for a service of benchmarking buildings for preparation of an organization sustainability report for an inspection event. Some services can be also internally invoked, for example, if a fault of a sensor is detected, an internal service of re-running a learning algorithm may be internally triggered. Once a schedule of service deployment is generated, then we use planning to prepare the optimal configuration for deployment. This minimizes manual work and reduces the wasting of resource as well as the risk of human error.

5.1 Scenarios of Cloud Resources Optimization for an Energy Company

Let us now consider two scenarios of a company providing a smart energy system for energy monitoring and control of energy-consuming devices. This company provides to its clients several types of cloud-based services: energy monitoring, energy dashboards, smart lighting, smart appliances, smart workstations and smart heating and cooling control. Moreover, it offers advanced analytics services such as building performance benchmarking, consumption prediction, equipment fault detection and predictive maintenance, the generation of various reports, etc.

These services are requested by clients and provided by the energy company on demand, in the same manner as cloud services. The services may be requested at different points in time and may be used for varying periods. Customers want to have flexible services that are available when they need them, for the period they need them. That way, they use service(s) only for the time the service is really necessary, instead of having the service for a fixed period of time.

However, this flexibility introduces uncertainty from the service-providing company side. It is difficult to know how many services will be requested as clients may request any number of services at any moment. Considering that the company may have certain financial limitations for the infrastructure, as well as that client requests may have different priorities, the optimization of cloud utilization starts to play an important role. The optimization of use of cloud resources ensures that the resources that do not have to be used are not used. This leads to both monetary and environmental savings. When the system scales from one building to more buildings, and eventually to a whole city (i.e., a smart city), savings also become financially significant.

Therefore, the two scenarios in which the energy company could optimize the use of cloud resources are: (1) after receiving resource requests from clients, the scheduling algorithm is used to generate of a near-optimal schedule of services to be deployed, and (2) once that the deployment schedule is in place, a planning algorithm is used to make a list of actions to be done (a plan) for an underlying deployment of dependent back-end services. In Figure 5.1, we present the system architecture, how cloud services for deployment are being scheduled, and thereafter how planning is used to deploy previously scheduled cloud services.

5.1.1 Scheduling of Cloud Services

Let's now turn to see how scheduling technique can be used to optimize cloud resources used for a smart energy system. To show the feasibility of the approach, a prototype has been implemented. This is a result of a joint work with Viktoriya Degeler and the related implementation is performed by her, while the problem description, the system architecture and the demonstrating example are provided by the author of this thesis.

In the following, we describe cloud resources and the specifics of requests for the cloud resources, the behavior of the system, and the responsibilities and functionality of each component of the system.

Cloud Resources

A cloud can be seen as a set of computational and storage resources. We refer to the smallest unit of a cloud infrastructure as a *resource*, be it CPU, memory or a single, stand-alone server. The term *Resource* is an abstraction that represents an instance of a cloud used as computing or storage capacity. Resources can be shared or exclusively used, and can have different services deployed onto them. All resources in a cloud that are available for usage are showed in a resource inventory. We assume that the number of available resources is limited by company's planned budget for the infrastructure. Sometimes, companies may also want to limit a number of used resources to a number of free cloud resources offered by cloud resource providers. We focus on resource scheduling, and, without loss of generality, we will use a hardware agnostic approach.

Requests for Resources

The request for resource utilization come from *resource requesters* (e.g., clients using a specific service). To request the resource, one needs to know which services are needed to fulfil a complete functionality or a job (e.g., reducing lighting energy consumption using a smart controller service and evaluating savings using an analytics dashboard), what type of resources are required, and for how long and in which way those resources will be used. These parameters represent input parameters for the scheduling service.

A *request for resources* is composed of the three following elements: a resource demand, a policy and a request. A resource demand contains: the resource type, the required number of resources, and information whether the resource can be shared or it must be exclusively used (e.g., a resource requester needs two instances of an Energy Dashboard service that may be shared or invoked by more service consumers). There can be more resource demands under one request for resources. A policy defines an amount of time for which a resource is required, and a parameter which defines how the resource needs to be used (e.g., the Energy Dashboard is used for five consecutive days without interruption). From the perspective of a client evaluating a Smart Control service, one request for resources could be: in order to evaluate the Smart Control Service which may be evaluated by Energy Dashboard Service, we need to demand two resources of types Smart Control Service and Energy Dashboard Service, which will be used exclusively for five consecutive days. This request for resources defines a dependency between two services.

System Behaviour

In Fig. 5.1, we show our proposal for a system architecture to provide cloud resources to resource requesters, taking into account the above-mentioned limitations. The sequence of actions and flow of information is the following. The resource requesters submit their requests to the *Resource Request Service*. Then, the *Scheduler Service* is invoked. The *Scheduler Service* provides an optimized schedule as an output. The schedule defines which resources are assigned to which time slot. Next, the *Planner Service* generates a list of actions (a plan) to deploy underlying back-end applications/services. When the *Deployer Service* receives a plan, it physically deploys the services to the resources, as per the plan. After the deployment process is finalized, testing scripts are executed in order to define the status of the services, or/and to execute the initial preparations of the services.

When services are up and running, the *Distributed Configuration Service* keeps track of the physical locations (endpoints) of the services, and gives an input to *Monitoring Service* which shows the current status of each individual deployed service. In case additional requests are submitted, re-scheduling can be dynamically invoked, while a number of used cloud resources would stay within the limitations given by the resource requester.

In the following, we explain each component of the architecture in more detail. The *Resource Request Service* is responsible for communication with the Scheduler Service and the preparation of requests in a form understandable by the Scheduler Service itself. The *Scheduler Service* is responsible for the provisioning of an optimized *Schedule* as the output for structured requests for resources

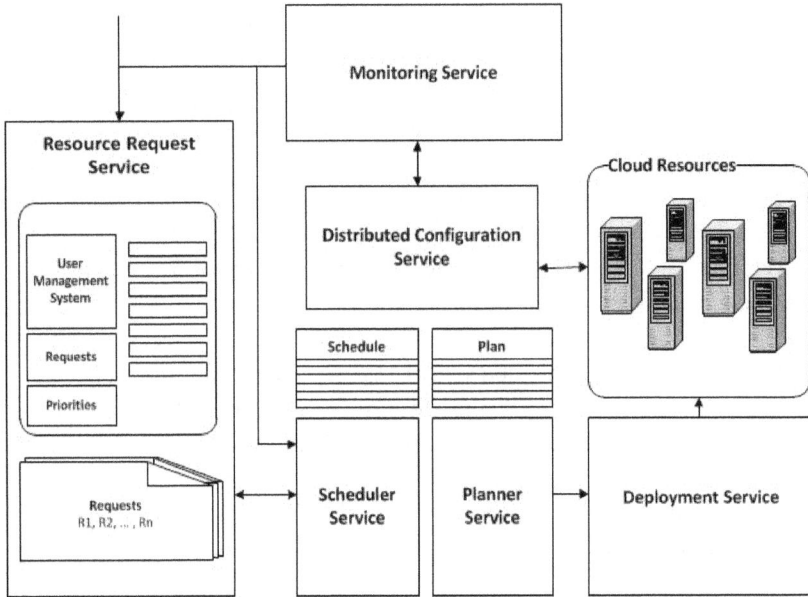

Figure 5.1: System Architecture

as an input. The provided schedule maps the requests for resources to the available time slots in an optimal manner. In the following section, the scheduler service will be described in more detail. The *Planner Service* is responsible for the creation of a plan necessary for the composition of applications or services to be deployed. The *Deployment Service* is responsible for the physical deployment of requested services to appropriate cloud instances. The input to the deployment service is a previously generated plan for services. The goal of the *Deployment Service* is to deploy the services and configure them to run. This process is improved by using the planning service, which is described in Section 5.1.2.

Additionally the *Deployment Service* is responsible for updating the *Distributed Configuration Service* (DCS). The DCS is responsible for the maintenance of information about the physical location of each cloud instance. Initially, when services on instances are deployed, it will send the update to DCS, which will store the information about its location. Location information will be represented in a form of endpoints that will point to specific instances. The *Monitoring Service* is responsible to represent the current state of each service deployed on a cloud instance (up and running, down, instantiating, deploying, restarting, etc.). Additionally, information on the performance of individual services is being collected.

It is worth noting that there are plenty of build automation tools that support different part of the described process. They support the build automation process by generating scripts, or by supporting continuous integration or configuration management processes. However, they depend on human factor for scheduling of services to be deployed or for configuration to be defined. In this work, we automate these processes to reduce human error and to optimize resource use.

Demonstrating Example

In order to show how Scheduler Service provides optimization, we have defined an example that represents a typical situation for a company providing energy consumption optimization services. This example is an equivalent to the example presented in (Nizamic et al. 2012) with a different application area.

Let us assume that this company provides to its clients several types of cloud-based services, e.g. energy monitoring, energy dashboards, smart lighting, smart appliances, and advanced analytics services.

To complete a specific task, quite often clients need a set of services to run together. For example, to evaluate an effect of the Smart Lighting Service and the Smart Workstation Service, a client needs also the Energy Dashboard Service. These services may come in different versions, e.g., the Dashboards having different data presentation on the front-end, and the Smart Control services using different control algorithms. The set of these services provides the full functionality of a task that a client needs to complete. That way, one request embodies a specification of needed complete working environment that provides the full functionality of the system. Also, one request links more services and that way implicitly defines dependencies among them. List of the requests for resources required by several clients (below represented by Client-i, where i=1..6) is presented in Table 5.1.

Each of clients has specific needs for resources. Those needs are reflected in the description how and how long resources will be used; if resources can be shared with other teams or not, and if they need to be used continuously, repeatedly or if some other policy should be implemented.

Given the total of 120 hours (5 working days 24 hours each), and the limit of maximum 25 simultaneously used resources, the schedule for one working week produced by the Scheduler is shown in the Table 5.2.

Total number of used resources provided by this schedule is 1680 server-hours, and it is optimal in regard to number of resources used in one working week. Every other schedule would lead to less or in best case equally good solution.

Table 5.1: List of requests

Request	Client-1		Client-2		Client-3		Client-4		Client-5		Client-6	
Shared?	Y	N	Y	N	Y	N	Y	N	Y	N	Y	N
SmartLighting v1						8		8	4			
SmartLighting v2		4	4							4		
Dashboard v1					4		4	1				
Dashboard v2	2		2									
Dashboard v3												2
SmartWorkstations v1							2					
SmartWorkstations v2	1		1		2						1	
Duration (hours)	72		2		48		24		2		24	
Cycle duration	-		24		-		-		-		-	
Number of jobs	-		-		-		-		6		-	
Policy	Cont.		Repeat		Cont.		Cont.		Multi		Total	

Table 5.2: Optimized schedule

	Mon	Tue	Wed	Thu	Fri
Client-1	20:00-23:59	00:00-23:59	00:00-23:59	00:00-19:59	
Client-2	22:00-23:59	22:00-23:59	22:00-23:59	22:00-23:59	22:00-23:59
Client-3			00:00-23:59	00:00-23:59	
Client-4					00:00-23:59
Client-5				08:00-19:59	
Client-6	20:00-23:59	12:00-23:59		20:00-23:59	20:00-23:59

Scheduler Service Performance

Finding the optimal schedule is an expensive task in terms of computational resources, known to be NP-hard problem (Chen et al. 1999). Practically, we ensure that the Scheduler can sustain a certain level of demand increase, and remain practical for higher number of requests.

As the scalability is an important characteristic of cloud computing, the performance evaluation investigates the ability of the scheduler to scale with the increase in the number of time slots, and also shows the usability of the scheduler with the increase in the number of requests.

The experiment was conducted on a 64-bit Windows 7 workstation (standard workstation provided by the university), comprising processor Intel Core 2 Duo CPU E8400 at 3.00GHz and memory 4,00GB RAM.

Number of Time Slots

Often, the scheduling of resources in a cloud is done on an hourly basis. Thus, if we take into account a working week, the number of time slots can be up to 40 (8 hours times 5 days). Thus the ability of the scheduler to scale with respect to time slots is important. Based on a typical scenario, we performed an experiment to run the Scheduler with 5 randomly generated requests and schedule them on a period from 5 to 50 time slots, representing a typical client request duration. The results are presented in Figure 5.2. As can be seen, even the scheduling for 50 time slots takes only about 2.8 seconds. Taking into account that this scheduling is done for the distribution of resources over a full week, we consider the performance to be within acceptable bounds.

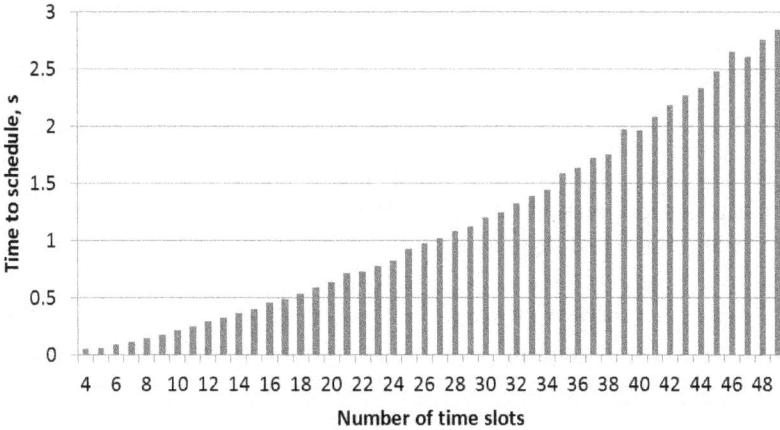

Figure 5.2: Scheduler performance based on the number of time slots.

Number of Requests

The number of requests causes much bigger strain on a scheduler, because at each time slot it needs to regard 2^{nReq} possibilities. As mentioned before, the scheduler is optimized to work well and to find optimal solution under small and stable number of requests. However, since we assume the possibility of requests increase, we implemented the dynamic relaxation of the optimality requirement, and instead we try to search fast for a solution that is satisfactory for a typical number of customer requests (i.e., 1-4 requests). The dynamic relaxation is done by implementing a gradual approach in the following way: if the number of requests is higher than a certain predefined number, in the experiment it was set to

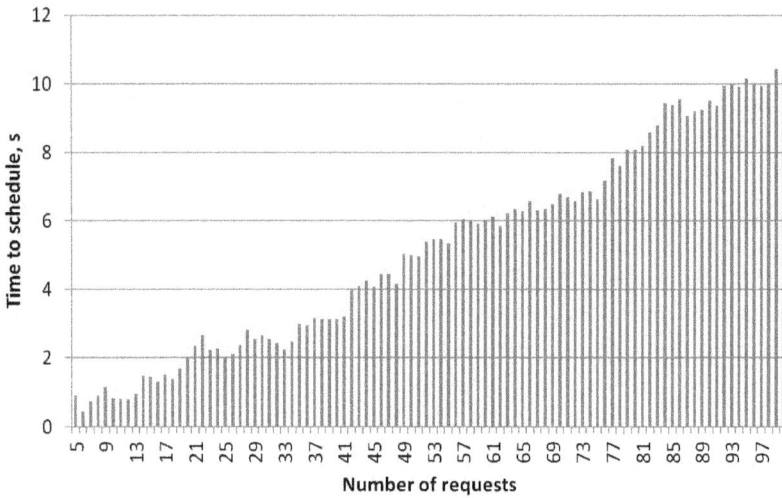

Figure 5.3: Scheduler performance based on the number of requests.

8 (i.e., double value of maximal expected typical number of customer requests), the requests are split on several groups. We run the Scheduler for the first group, obtain the optimal schedule for this group, and than freeze the already scheduled requests in their time slots, and begin to schedule a second group, taking into account the already scheduled requests, and so on. Note that while this approach is "greedy", thus not guaranteed to return the optimal solution for the full number of requests, the returned solution is still effective, because if the next group of requests contains resources that can be shared with those in the previous groups, this situation is always detected and it automatically gives preference to those time slots that allow for maximum sharing of resources with already scheduled requests. Figure 5.3 shows the time needed to schedule up to 100 requests.

5.1.2   Planning of Cloud Applications Deployment

Planning techniques can further improve the deployment process. To show the feasibility of the approach, a prototype has been implemented. The prototype is a result of a joint work with Ilche Georgievski and the related implementation is performed by him, while the domain knowledge and demonstrating example are provided by the author of this thesis.

In the following, we: (1) describe cloud applications for a smart-energy system, (2) describe deployment as a planning problem, (3) present a demonstrating example, (4) present hierarchical planning domain model and finally (5) present experimental evaluation.

Deploying Smart-energy Applications in Cloud using HTN Planning

Cloud services are provided by deploying applications to a cloud. The applications are no longer installed and run on a single machine, but they are composed of assorted software components that are deployed and distributed across machines of cloud infrastructures. Given some initial configuration of a cloud infrastructure in terms of already deployed components, a set of deployment actions (such as start and stop a component), a desired application, a deployment problem consists of finding a sequence of deployment actions over components that compose the desired application.

While the deploying aspect of this process is already fully automated, the composition of application components is yet to be improved, being still performed either manually or semi automatically with some predefined scripts. The scripts may ease the process to a certain degree, but their use is limited as they are exclusively dedicated to specific components and applications. Moreover, even a small number of components can make the composition process already strenuous and difficult. Thus, the difficulty increases further with the number of components to be configured, especially when the components are delivered over several builds and releases with different compatibilities among each other. The process of satisfying interdependencies between components then can no longer be performed manually or with the mainstream tools.

The following is a concrete case from the previously presented domain of smart-energy applications. Such an applications generally consist of a number of primary components responsible for dealing with core processes (*e.g.*, sensing and processing information, executing actions), and several secondary components that complete the life cycle of the whole system (*e.g.*, providing communication means, storing the raw and context data). While being highly interrelated with each other, these components are implemented as cloud services, and each one can have multiple versions. Their composition is to be deployed on a cloud with a cross-platform environment. In actual deployments, the number of components that need to be composed can increase as some components may have multiple instances running, for example, to cover different spaces (*e.g.*, floors, offices, common spaces, *etc*). This increase in the number of versions and instances to be deployed, as well as the dependencies between them, make the deployment problem highly challenging.

Approach

A common means to facilitate composing components is by using automated planning. Various planning techniques are already well studied for composing, for example, Web services (Sirin et al. 2004, Sohrabi et al. 2006, Kaldeli et al. 2011) and information flows (Riabov and Liu 2005, Sohrabi et al. 2013). In the context of cloud applications, the use of planning, especially general-purpose one, is scarce. Therefore, we introduce an approach based on general-purpose planning. We use Hierarchical Task Network (HTN) planning to address the problem of automated composition of application components. HTN planning is particularly useful due to its rich domain knowledge and advice on how to accomplish something. Such knowledge helps in reducing the search space and therefore finding a solution reasonably fast, if one exists. We propose a strategy to create an HTN planning problem from a deployment problem. We use the so-called Aeolus model (Di Cosmo et al. 2012) to define the deployment problem. In this model, components are resources of various kinds that require and provide functionalities through ports. Requiring and providing functionalities implies establishment of interdependencies between components. A requested application is realised by a sequence of low-level actions, such as create instance, start instance, bind port, and so forth. We demonstrate the applicability and feasibility of this approach through an experimental evaluation, and we show that general-purpose planners can be linked to specialised ones to a certain degree. We use the state-based HTN planner SH (Georgievski et al. 2013) for the implementation and evaluation of our proposed solution.

HTN planning provides the means for solving deployment problems, and the Aeolus model enables specifying them. HTN planning (Erol et al. 1994) is popular and well suited for various domains due to its rich domain knowledge (Georgievski and Aiello 2015). The domain consists of tasks that can be accomplished by operators or methods. An operator represents a transition from a state to another one, while a method predefines how to decompose some task into greater details. Given an HTN planning problem, which consists of an initial state, an initial task network and sets of operators and methods, planning is performed by repeatedly decomposing tasks from the initial task network until operators executable in the initial state are reached.

Deployment Model

One way to define the problem of configuring and deploying applications on the cloud is by using the Aelous model (Di Cosmo et al. 2012). The main element of the model is a *component*, describing a manageable resource that provides and

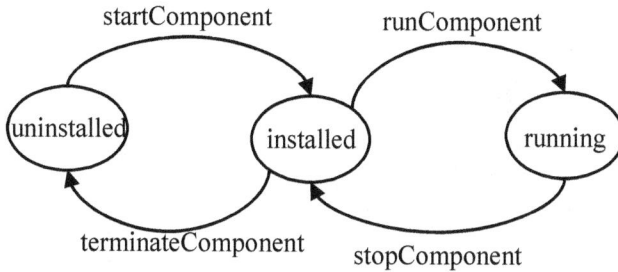

Figure 5.4: Finite State Machine depicting the state transitions of a component

requires functionalities. Through the use of state machines, the Aeolus model provides a way to encode specific components declaratively by specifying how functionalities are accomplished. The declarative form makes the model particularly suitable for a wide range of AI planning techniques. The Aeolus model allows for a finite set of *states* for each component. Since the number of states in practice is rarely higher than three, for simplicity, let us consider a component as the Finite State Machine (FSM) shown in Figure 5.4. The FSM defines the *state transition* processes of a component. A component is initially in an *unintalled* state. Upon start, it transitions into an *installed* state, and then to a *running* state. State transitions are accomplished using *deployment actions*. For example, given some component in its initial state, it is installed by invoking the *startComponent* action. A component that is in a running state can be stopped using *stopComponent*, while a component can be eliminated from the current context by invoking the *terminateComponent* action.

A deployment problem consists of an initial configuration, a set of deployment actions, and a request for a new configuration (*i.e.*, application). The solution to the problem is a deployment run representing a sequence of deployment actions on components that, when deployed, produce the required configuration.

Deployment as an HTN planning Problem

We introduce an approach to create an HTN planning problem from a deployment problem. We provide a demonstrating example that helps in demonstrating the structures we encode in the domain model. We use the Hierarchical Planning Definition Language (HPDL) (Fdez-Olivares et al. 2006) when describing the planning structures. In the following, we refer to a state transition that does not depend on any functionality provided by other components as *simple transition*. Otherwise, we use the term *complex transition*.

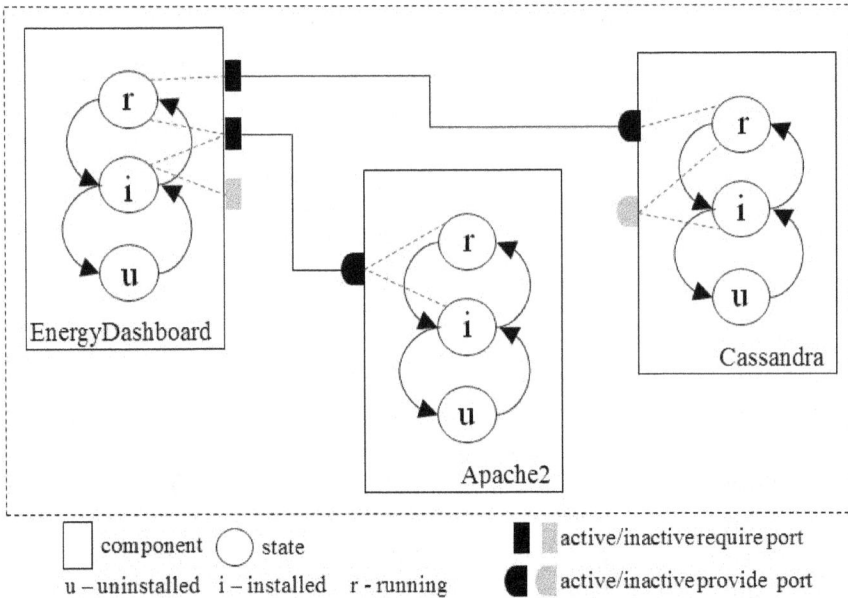

Figure 5.5: Example of a pattern for an *EnergyDashboard* application

A Demonstrating Example

Figure 5.5 graphically represents an Aeolus pattern for composing an EnergyDashboard application in a running state. The main and top-level component
represents *EnergyDashboard*, which is a dashboard displaying energy consumption. *EnergyDashbboard* operates using several software services among which essential ones are a Web server and a database. The application requires a database
to retrieve all consumption information (*e.g.*, total building consumption, individual user consumption). In this case, *Cassandra* and *Apache2* as components
that *EnergyDashboard* depends on.

Hierarchical planning domain model

We describe a strategy to encode elements and features of the Aeolus model
mainly into operators and tasks in an HTN domain model.

We   encode   components,   instances,   ports   as   domain   types
`component instance port`, which are all subtypes of the type `object`. In
fact, each component type, such as *EnergyDashboard* is represented as an object

of type `component`.

While FSMs associate components with states abstractly, component instances are the ones to be in a specific state at planning time. We encode an instance state using a predicate "*(state instance)*", where *state* is a string representing the type of an FSM state, and *instance* is a variable representing the component instance. An example of a *EnergyDashboard* instance *w1* in an installed state is `(installed w1)`.

A component state may be associated with require and provide ports. To represent the association of a port to a state, we use a predicate "*(statePort component port)*", where *statePort* is a string representing the type of port in a specific state, *component* is a variable representing the type of component that requires or provides a port represented by the variable *port*. For example, if *EnergyDashboard* requires the *httpd* port in the installed state, we encode it as `(installed-require energydashboard httpd)`. Such knowledge holds for all instances of the respective component. These predicates are therefore grounded in the initial state and static during planning.

Configuration Processes

Although each different type of an application has its own installation and running configuration pattern, the process of configuring applications is general and can be abstracted away. Let us detail how we can accomplish that.

The process of configuring an application requires satisfaction of the dependencies to functionalities provided by components. Let us assume that an instance in an uninstalled state cannot have requirements to be satisfied. We may then consider two abstractions for complex transitions of components. The first abstraction refers to acquiring a component functionality in the installed state, while the second one refers to establishing a functionality in the running state. We point out that complex transitions representing other configuration types can be easily incorporated in the current domain model with minor modifications. HTNs naturally enable encoding knowledge at different levels of abstraction. This support for modularity enables us to focus on a particular level at a time (Georgievski and Aiello 2015). We can formulate tasks and encode high-level strategies in the methods of these tasks before reasoning on low-level tasks (operators).

We encode each abstraction as a task in the domain model, namely `install` and `run` tasks. Each method of these tasks encodes a specific case. One such method involves port activation. If a component state is associated with one or more require ports, the *port activation* process makes sure that the need of the current instance for specific functionalities is addressed. That is, if the current

component instance has require ports that are not active, the method first activates each port and calls recursively its corresponding task until all necessary
ports are activated. The actual process of port activation is encoded in a separate task. The task not only activates a required functionality, but also finds
and installs (or runs) a component instance that provides that functionality. An
instance with active require ports can then use the functionalities of other components with active provide ports. This is accomplished by another method that
involves port binding. The process of *port binding* binds require ports to appropriate provide ports. For this process, the method depends directly on the
binding actions. Once we have methods that involve port activation and binding, we can proceed to the method that deals with the case when all require ports
are active and bound. To address the satisfaction of all require ports, we use a
forall expression in the method for both tasks, install and run. The following
expression is used for the install task.

```
(forall (?p - port)
        (and (installed-require ?c ?p)
             (bound ?p ?i ?i1)))))
```

After this constraint check, we are ready to start or run an instance. In the case of
the run task, when running an instance, we have to deactivate the ports that will
be no longer provided by the instance in the installed state. The process of *port
deactivation* is accomplished using a separate task with multiple methods. Each
method represents a different case to be handled, such as a provide port that is
bound but needed for the running state, a provide port free to be unbound, *etc.*
The port deactivation task uses port unbinding. The process of *port unbinding* is
more complex than the binding one, and requires checking for constraint viola-
tion. That is, we have to take care of active provide ports bound to active require
ports. We use a separate task for this process, that is, unbindPorts. This task
does nothing when the port is bound and needed for the next transition. When
all necessary constraints are satisfied, it unbinds a specific port and recursively
calls itself, shown in the following encoding. Being a recursive task, it includes a
base case that performs phantomisation (Georgievski and Aiello 2015).

```
:tasks (sequence (unbind ?p ?i ?i1)
                 (unbindPorts ?i))
```

There are methods in the install and run tasks that deal with the case
when there are no required functionalities for an instance. This means that we
have a simple transition which can be handled by installing the component instance directly. In the case of running an instance, we invoke the port deactivation task to ensure a valid transition to the running state.

The modelling of the transitions from a running state to an installed state and further to an uninstalled state is analogous to the encoding of the tasks we described so far.

One of the features of these kinds of compositions is that a cycle may occur between states of different component instances. That is, an instance is expected to provide a functionality at a specific point in the composition, but it is not possible because at the same point the instance is required to change its state (Lascu et al. 2013). We address this feature using the process of *instance duplication*. Instance duplication deals with such cycles by creating as many instances of the same component as needed, and deploying them in different states at the same time. We encode instance duplication as a separate method. The method makes sure that the current component instance is in a specific state and it has at least one provide port bound. Consequently, a new component instance is created either in an installed state or in a running state, depending on the type of configuration.

Experimental Evaluation

The main objective of our experimentation is to evaluate the applicability of our approach for composing cloud applications. To address our objectives, we generate deployment problems of increasing number of components representing a typical application presented in the demonstrating example, varying from 3 to 220 components, resulting in more than 50 problems. We apply our approach to create the corresponding HTN planning problems, and examine the performance of SH planner component (Georgievski et al. 2013) on them. The HTN planning problems are constructed from deployment problems of varying difficulty. For example, the difficulty of a problem can be increased if there is a need for instance duplication. Also, for SH, an HTN problem can be more difficult if the requested configuration appears deeply in the search space. To that end, we construct two cases of deployment problems mainly following the test pattern provided in (Lascu et al. 2013). For both test cases, we use a set of components $c_1, \ldots, c_n$, where each c_i has require and provide ports as follows. Given that we want to have the rightmost component c_n in its running state, the dependencies between components will require to first create instances for components from c_1 to c_n, then to perform transition from uninstalled to installed state in the reverse order of component instances, and finally, to transition from installed to running state in the order from c_1 to c_n. We modify the second test case in such a way to require instance duplication. In particular, we randomly select several components and, for a selected component c_i, we remove the activation of a provide port p_i^1 from its running state. The removal requires another instance of c_i to be

Without duplication			With duplication		
Problem	Plan length	Time (sec)	Problem	Plan length	Time (sec)
3	12	0.077	3	16	0.017
6	27	0.032	6	35	0.004
10	47	0.041	10	55	0.012
20	97	0.193	20	109	0.046
30	147	0.226	30	171	0.113
50	247	0.354	50	287	0.389
70	347	0.784	70	399	0.898
100	497	1.957	100	577	2.34
120	597	3.112	120	693	3.871
150	747	5.791	150	863	7.182
180	897	9.625	180	1037	11.916
200	997	12.897	200	-	OM
220	1097	16.918	220	-	OM

Table 5.3: Evaluating the applicability of our approach by using SH planner under increasing problem difficulty ("OM" signifies "out of memory")

created so as to satisfy the requirements of c_{i-1} and c_{i+1}.

We show a subset of our results in Table 5.3. The left-hand side of Table 5.3 shows the results of the first test case without instance duplication, while the right-hand side shows the results with instance duplication. Columns three and six show the time in seconds needed to find a solution. For each problem, we show the plan length as an indication of the difference between the number of operators creating instances and the number of other deployment actions. In the case without duplication, the number of generated instances equates to the number of components, while in the latter case it is strictly greater than the number of components. With the creation of a new instance, we increase the size of the state by adding two predicates, and modify the state by updating the domain function.

All problems are solved within 17 seconds. When the number of components is larger than 120, the need for instance duplication degrades the performance of the SH planner as compared to the case without instance duplication. However, in typical applications, there are no too many scenarios with more than 100 components for which case SH planner can find a solution in about 2 seconds with and without instance duplication. The results also show that the planner runs out of memory when the problem has more than 200 components with instance duplication and 220 components without duplication. This is mainly due to the implementation of the core part of SH, which employs recursion (in these test cases, the number of recursive calls increases rapidly), and the need for creation and maintenance of a large set of objects.

These results also address our second objective and show that general-purpose planners can exhibit a satisfactory performance in composing cloud applications. Compared with the results of the two general-purpose planners reported in (Lascu et al. 2013), our HTN planner outperforms both planners significantly. Compared with the specialised planner, our planner falls behind the specialised one only after a high number of components. This implies that other planners should be examined and evaluated in case that number of components of a typical application increases.

Chapter 6

Evaluation

We evaluated all deployed solutions from the perspective of environmental and economic savings. In the context of the present thesis, it is important to define what we mean by environmental savings. As the implemented changes relate to electricity, water, and waste, that directly or indirectly relate to environmental footprint, we refer to our achieved savings as environmental savings.

To evaluate and compare solutions or interventions, we define measures that enable us to compare. More specifically, each solution that relates to electricity savings is presented in *kWh/year*. The solution that relates to waste reduction is expressed in *kg/year*, and, finally, water savings in $m^3/year$ [1]. To compare solutions which achieve savings for different types of utilities (e.g., electricity and water), we present the increase in efficiency as *percentage (%)* of savings introduced with newly installed systems. Increase in efficiency represents the difference between baseline measures and measures after solutions have been applied. This also represents absolute measures that can be used to compare project savings and understand potential impact if a solution is scaled to a larger space or ported to another similar space.

Additionally, for economic savings, measures of savings are presented in *EUR/year* per deployed solution, as well as *in years* for payback periods, and in percentage *(%)* for the return on investment (ROI).

6.1 Environmental Savings

All the conducted experiments have as goal to investigate what additional savings can be achieved in addition to an existing building management system in the Bernoulliborg. The savings represent additional optimization and show how much less electricity, water or waste can be used if the system is in place. We present below all saving solutions, the experimental setup in the real environment, and the results of the experiments.

---

[1] 1 m^3 (cubic metre) = 1000 litres

6.1.1 Lighting Control

As previously presented in Section 4.3.3, the function of the Lighting Control system is to control lights based on environmental data collected by sensors (e.g., PIR and LUX sensors), with the goal of reducing energy consumption whenever and wherever possible.

We chose the restaurant as an experimental environment. It is located on the ground floor of the Bernoulliborg. The restaurant covers a total area of 251,50 m^2 and has a capacity of 200 sitting places. The restaurant has glass walls on three sides, providing a significant amount of natural light when weather conditions allow. The restaurant area is used for lunch in the period from 11:30 a.m. 2:00 p.m. Outside these hours, the area is used by staff, students, or other visitors for working, meeting, or social activities.

The restaurant area is an open space divided into two sections by construction. We make use of this division in the experiment. In particular, each section has 15 controllable light fixtures, making 30 in total. There are several light fixtures that are uncontrollable; these are security lamps. While we do not control these, we take into account the light that they provide. In addition, there are two types of controllable fixtures. The first are large and have 38W of power consumption each, and the others are small, each having 18W. These fixtures are controllable by using the actuators attached to them, which also serve as sensors by providing information about the fixture's power consumption. We installed 15 more sensors, one to measure the natural light level, and the others to detect people's movement. In order to make more meaningful use of the restaurant space given the movement sensors, we divided each section into smaller spaces, called *areas*. In our case, a section was comprised of multiple areas. In each area, we embedded a movement sensor to understand environmental conditions (e.g., light levels, and presence or absence of occupants within the area).

To define a baseline, over the course of two weeks in February we recorded measures of state (on or off) and power consumption (W) of light fixtures in order to understand the typical behavior of manual control of lamps in the restaurant. Data about the state and consumption of each controllable lamp was collected for each lamp individually using smart plug measuring devices. Frequency of consumption data collection was five minutes. State data was collected whenever the state of a device changed (e.g., from ON to OFF). Measured values were transferred to our database using the deployed system. The state data was reliable, while the consumption data lacked some values due to smart plug errors. Having reliable state data enabled accurate calculation of the time when lights were turned on. By knowing exact consumption per lamp and its status for each second, we calculated the consumption of each lamp.

Observing the measurements gathered in February 2015, when the restaurant was controlled manually, we found that the average time point when the lamps were turned on was 6:30 a.m., and they stayed on until 8 p.m., which is also an average point in time. The measurements show that in the baseline collection period (from Feb 2 at 00:00 until Feb 15 at 23:59:59), total energy consumption was 130,68 kWh. This meant that the average baseline consumption per working day in the restaurant was 13,5 kWh. For weekends there was no manual control of the lamps, thus no consumption. When normalized per day, average consumption in the baseline collection period was 9,33 kWh/day. These numbers were used as baseline consumption for later comparison and identification of efficiency.

We conducted the experiment on the system over the course of several weeks in March and April 2015, involving measurements from Monday to Sunday. In the last week of March and first week of April, we allowed our system to control the environment in order to reduce electricity consumption of the Bernoulliborg restaurant. Thus, manual control of lights was disabled, automated control of lights was enabled, and the system was running continuously without interruptions during these two weeks. As follows, we present the benefits of the system in terms of the energy savings resulting from the coordination of light fixtures to correspond to weather conditions, presence of people and a set of minimum requirements for satisfactory light levels. As there were plenty of possibilities for lamps to be turned off, this solution provided energy saving. Figure 6.1 shows the average energy consumption when the lamps in the restaurant were manually controlled and when our automated Lighting control system was used. In contrast to the manual control, which assumes almost fixed time points for turning on/off lamps, our system reduced the consumption of these lamps by turning them on only when actually needed. Figure 6.2 shows the automated use of lamps and therefore energy on each day. The top part refers to the first week of using the automated system (depicted with a full line), and the bottom one depicts the results for the second week. We also include the estimations of consumption if manual control had been used (depicted with dashes). In addition, the figure includes weekends when there is no regularly provided manual control.

The measurements show that in the automated control period (from March 25 at 00:00:00 to April 5 at 23:59:59) total energy consumption was 17,41 kWh. When normalized per day, average consumption in the automated control period was only 1,45 kWh/day.

For this experiment, data on external lighting conditions was used for auto-

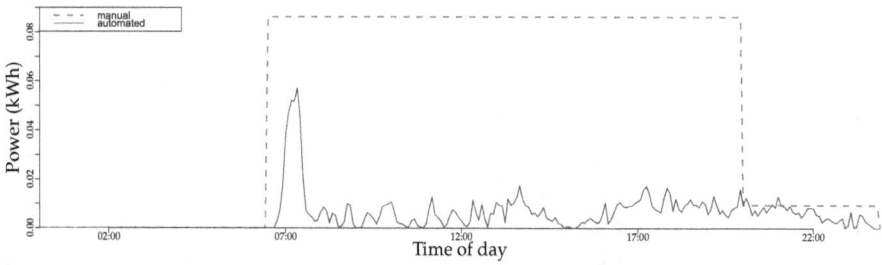

Figure 6.1: Comparison of average energy consumption between manual control and control by our system

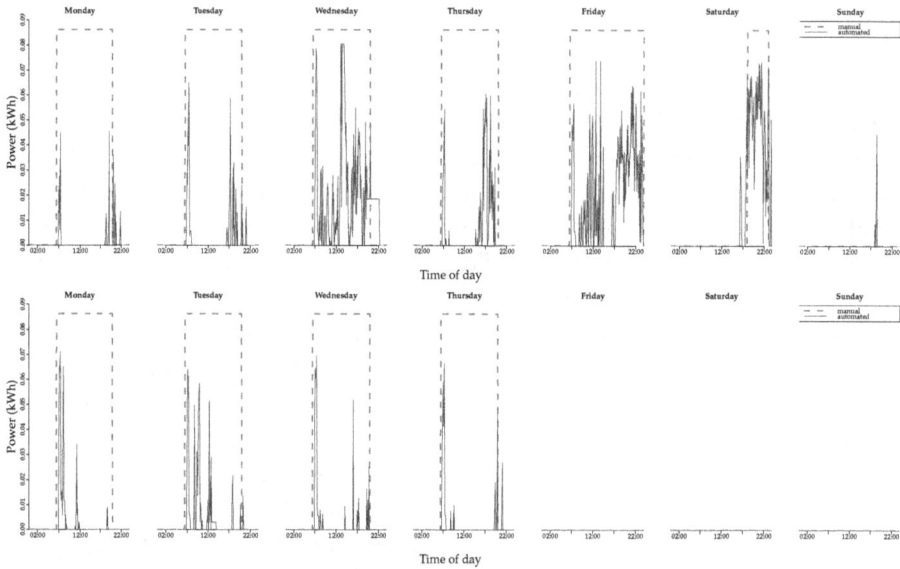

Figure 6.2: Comparison of energy consumption between manual control and control by our system for each day during week one (upper charts) and week two (bottom charts)

mated control, but not stored in the database. However, weather data shows that both periods had very similar weather conditions. The average savings of energy between the scenario of manual control and the one with our system is in order of **80%**.

The high efficiency of this solution is supported by the physical characteristics of the restaurant (i.e., three walls bordering with the outside space are made of windows) and initial usage pattern, i.e., the restaurant is effectively used only for about 2 hours during lunch, whereas the lights are turned on for an average of 13,5 hours. As we understand, the lights in the restaurant are on based on faculty policy and the time when this space is expected to be used by students, staff and visitors (i.e., also for purposes other than lunch, such as studying or meetings).

The fact that the baseline data collection and experiment were done with a break in between does not affect the results. This is because the baseline is always the same, taking into account fixed scheduled manual control of all the lights in the restaurant. Moreover, the experiment was done without interruptions during two consecutive weeks in the non-exam period, which implies stable occupancy of the space.

Though the presence of people in the restaurant at evenings and during weekends was rare, there were still special occasions when our system encountered them (see Friday evening and Saturday on the top part in Figure 6.2). This demonstrates that our system made the restaurant truly adaptable to the happenings within. To have a fair comparison, we assumed that on special occasions, such as dinner for Faculty guests or excursions through our living lab, there would be manual control of the lamps provided in the restaurant. Also, one may note that the Friday in the second week was a special occasion, a holiday.

Working days	5
Working weeks	51
Number of light fixtures	30
Average time light fixtures are ON	13,50
Average power consumption per light fixture [W]	34,66
Annual energy consumption for the restaurant [kWh]	3.580
Measured savings by using the system [%]	86
Projected annual savings [kWh]	3.078

Table 6.1: Projected annual savings for the automated Lighting control in the Bernoulliborg restaurant

Taking into account the annual energy consumption for the restaurant and the measured savings by using the system, we projected the annual savings, as shown in Table 6.1. The annual savings for the Lighting control solution were **3.078 kWh/year**, while the efficiency increase was in the order of **80%**.

6.1.2   Computer Sleep Mode Control

As previously presented in Section 4.3.4, the Computer Sleep Mode Control has as its function to gather data on the status of computers and their usage, with a goal to adjust sleep mode timeout value to minimum user-satisfying value to introduce additional energy savings from computers. To show the environmental savings from the Computer Sleep Mode control, several experiments were conducted. This was a joint work with Brian Setz, and he carried out the experiments and data collection during work on (Setz 2015).

Together with the system administrators of the University of Groningen, we identified fourteen computers on which to conduct an experiment before expanding the solution to the whole building and potentially to the whole University. The PCs were connected to our system using the same local area network.

The data set used in this research was obtained by using the Computer Sleep Mode solution to monitor computers for an extended period of time, as well as by collecting inputs from users through a survey (Appendix 1). The PCs in question were the PCs used by employees at the University of Groningen who had volunteered to be part of the experiment. The time period during which this data was collected was between two and three months, depending on the individual computer. The minimum timespan needed for learning the user behavior was one month. One month was needed, as profiles were built for each computer, each profile requiring at least four weeks of historical data collection.

For each computer four different types of data were collected: state data, activity data, feedback data and timeout data. The state data refers to the state of the PC (on, off, sleeping), the activity data refers to activity on how the computer is used (active, idle), and the feedback data refers to feedback received from user behavior (e.g., if the user stated that the sleep mode was being activated faster than desired). Finally, timeout data relates to data about sleep mode timeout (e.g., the default sleep mode setting for Windows 7 is 30 minutes). The data about the state of the PCs was collected whenever the state of a PC changed (e.g., from on to sleep). The collected data was reliable and there were no missing data points.

The PCs used in the Bernoulliborg are standard workstations provided by the university. We performed power consumption measurements and observed that the average consumption per PC was 120W. To determine the baseline, we used the average time PCs were used daily, being 5 hours/day. Estimated annual consumption was 135 kWh/year/PC.

Experiments were performed by Setz (2015) to determine the effectiveness of the three tested models for adjustment of the sleep timeout. The models tested were: (1) Activity Probability model, Activity Probability and Negative Feedback model, and (3) Activity Probability, Negative Feedback and Idle Time. The goal of the *Activity Probability* model is to always have a high sleep timeout whenever there is a chance of user activity at that given moment in time. The sleep timeout becomes low when there is no probability of user activity. This model is a rule-based model. The goal of the *Activity Probability and Negative Feedback* model is to find the optimal sleep timeout based on two input parameters: activity probability and negative feedback. Similarly, the goal of the *Activity Probability, Negative Feedback and Idle Time* model is to find the optimal sleep timeout based on three input parameters: activity probability, negative feedback and idle time. The last two models are linear regression models.

The experiments lasted three weeks, during which each computer under experiment used each of three tested models. The experiments started on the 1st of June, 2015 and ended on the 22nd of June, 2015. The results of the experiments were analyzed to determine which of the three models performs best, and how much energy can be saved. Using the most aggressive control model (Activity Probability, Negative Feedback and Idle Time), it was measured that a PC could be put into sleep mode for 9,87 hours per month (0,49 hours per day, assuming that PCs were turned off during weekends). As shown in Table 6.2, the annual savings per PC are calculated for 45 working weeks.[2]

Working days	5
Working weeks	45
Average measured power consumption of a PC while sleeping [W]	5
Average measured power consumption of a PC while working [W]	120
Estimated average time a PC is used daily [h]	5
Measured average time a PC can be in sleep mode daily [h]	0,49
Average hours when a PC is working in a year [h]	1.125
Average hours a PC can be in sleep mode in a year [h]	111,04
Annual consumption of a PC while working [kWh]	135
Annual saving per PC [kWh]	**12,77**
Annual saving per PC [%]	**9,46**

Table 6.2: Annual savings per PC

Savings were calculated by taking into account the estimated average daily time a PC is used, and the measured average daily time a PC can be in sleep mode. The annual saving was calculated as annual consumption in case a PC

---

[2]An average employee of the University of Groningen works for an average of 45 weeks per year.

would have been turned ON instead of being put to sleep by our solution. The efficiency of this solution was found to be **9,46%**.

It may be of interest to note that the results of these experiments are based on people who always turn their PC off after work. Thus they are already quite efficient to start with. Even though the savings of 12,77 kWh/PC/year do not seem to be high when compared to the total consumption of the building (1.350.000 kWh), taking into account that the whole organization has more than 300 work places with a PC, potential savings from PCs add up to **3.831 kWh/year**, which is almost equivalent to the annual consumption of one average Dutch household.[3] Considering the number of typical office buildings around the world, the number of computers used, and the potential savings per computer, if this solution were widely applied its environmental impact would be significant. Moreover, it would have been interesting to compare the results from staff member computers with the results from the publicly available student computers, as publicly available computers are often left running while not being actively used.

6.1.3 Sensor Holder

As previously presented in Section 4.4.1, the Sensor Holder solution enables adjustment of sensor direction with the goal of increasing presence detection and reduction of sensor timeout. This leads to energy use optimization by providing lights only when they are needed and for the period they are needed.

The main problem was fixed sensor positioning in offices, as their visibility range is limited and the location of workers changes with time. To increase presence and reduce sensor timeout, the ability to dynamically adjust sensor direction is rather important. This reduces the time when lights stay on unnecessarily.

In the offices of the Bernoulliborg, most of the PIR (movement) sensors were located far from the location where occupants were performing their work. For this reason, sensor timeouts were set to be longer than the office occupants needed, causing the lights to stay on unnecessarily for 30-45 minutes after the occupant had left the office.

To understand the size of potential optimization, we collected data about the sensor timeouts in each office. The data was collected manually by demounting sensors and visually reading the value. The readings were recorded using an online application. Almost all timeouts of PIR sensors were set to 45 minutes, while only a few were set to 30 minutes. In total, we examined 180 offices.

To optimize consumption in the offices, we developed a hardware solution-

---

[3]http://www.wec-indicators.enerdata.eu/household-electricity-use.html

sensor holder. The sensor holder makes it possible to reduce the timeouts of
PIR sensors from 45 to 15 minutes. This is possible because the holder enables
pointing of the movement sensors in the direction of the employees, which gives
better presence detection as well as opportunity to reduce the light timeout to a
minimum.

In the course of three weeks, sensor holders were deployed in all offices in the
Bernoulliborg. On average, there were 5 lights per office, with an average power
consumption of 44 W. Taking into account the hours that lights were turned on
before and after the intervention, as well as the fact that there are 45 weeks when
offices are used effectively, the annual savings were estimated.

	Offices	Lights	Cons. (W)	h/d	d/w	w/y	Cons./y (kWh)
Before	180	5	44	7,41	5	45	66.023
After	180	5	44	5,31	5	45	47.312
				Estimated savings:			18.711

Table 6.3: Projected annual savings from sensor holders change

We estimated that this intervention enabled reduction of time when the lights
were on from an average of 7,41 hours/day to 5,31 hours/day. This led to projected annual savings of **18.711 kWh/year** for the whole building, as presented in
Table 6.3. In other words, the efficiency of office lights was increased by **28,34%**.

6.1.4 Water Consumption Reduction

As previously presented in Section 4.4.2, the purpose of the Water Consumption Reduction solution was to collect data on water consumption, introduce
continuous monitoring and reporting, to provide understanding of how water is
consumed within a building as well how this consumption can be affected, both
by retrofitting as well as providing feedback to the end-users responsible for the
water consumption.

Water consumption reduction solution was motivated by the observation that
a considerable amount of water was unnecessarily wasted when building occupants used the faucets (e.g., for washing hands or dishes). Before the intervention, the average water consumption of the Bernoulliborg was calculated based
on data from 2013 and 2014, and it amounted to 3.021 m^3.[4]

To be able to determine baseline consumption and to measure savings, five
new water meters were installed. Next, water consumption was measured for

---

[4]The numbers were provided by the energy manager of the University of Groningen.

the first time in mid- January 2014, as shown in (Figure 6.4). The measurements were collected by visual readings from the water meters, and recorded using an online application.

Timestamp	Main meter reading	Water consumption (m^3)
1-13-2014 10:23:38	15.704	12
1-14-2014 10:52:06	15.716	13
1-15-2014 10:32:53	15.729	12
1-16-2014 10:26:42	15.741	12
1-17-2014 10:40:33	15.753	12
	Total:	**61**

Table 6.4: Measurements before the water reduction change

The data collected was reliable as it was collected and verified on each of the measurement days. After the first week of data collection, measurements showed that baseline consumption was 61 m^3.

To save water, water saving devices were installed on thirty-five water faucets (i.e., water taps) within the building. Water saving aerators were added to the faucets, reducing their water flow by 50% per faucet. It is important to mention that water consumption by water faucets represents only a portion of the total building water consumption. Most water is consumed for the needs of the restaurant (e.g., washing dishes) and toilets.

Timestamp	Main meter reading	Water consumption (m^3)
1-20-2014 10:27:18	15.765	12
1-21-2014 11:11:09	15.777	16
1-22-2014 10:24:17	15.793	10
1-23-2014 10:24:08	15.803	10
1-24-2014 10:40:22	15.813	10
1-27-2014 10:14:46	15.823	
	Total:	**58**

Table 6.5: Measurements after the water reduction change

The data collection was repeated after the installation of water flow reducers, at the end of January 2014, as shown in Figure 6.5. After the second week of data collection, measurements showed that baseline consumption was reduced by 3 m^3, making the reduced consumption to be 58 m^3. The aggregated results and percentage of savings are presented in Figure 6.6.

	Week total (before) [m^3]	Week total (after) [m^3]	Difference (%)
Water	61	58	-5

Table 6.6: Comparison before and after the water reduction change

The total of water savings for the whole building was **5%**. Taking into account the average water consumption of the Bernoulliborg, it is projected that by means of this intervention water savings could be in the order of **151 m^3/year**.

This percentage could be larger during exam periods when more students are in the building, and smaller during the vacation periods. Moreover, if wireless water meters were available, this process could be automated easily (Musters et al. 2014).

6.1.5   Waste Separation Process Change

As previously presented in Section 4.4.3, the purpose of the Waste Separation Process Change was to collect data on separated waste, to observe possible optimizations to be derived from changes in this process, as well as to propose a hardware and software system to support automation of this process.

Before this intervention, the waste separation process in the Bernoulliborg involved only separation of general waste and paper. Waste bins for general waste and paper were available in offices and common spaces. Once a day, cleaning personnel emptied the bins.

To understand the initial situation and to define the baseline, before the change was implemented the amounts of general waste and paper were measured. The baseline measurements took place at the end of September 2014. The amount of collected waste was measured at the collection point using an industrial scale. The collected data is presented the Tables 6.7, 6.8, and 6.9. The total amount of annually collected waste was not known by the building management, as waste was collected jointly for the Bernoulliborg and another neighboring building.

To ensure reliability of the measured data, the data was collected and verified by three persons. Baseline consumption shows that within one working week 313 kg of general waste and 68 kg of paper were collected.

Once the waste separation process had been completely changed within the building, instead of being able to dispose only general waste and paper the occupants of the Bernoulliborg were able to separate two additional types of waste: plastic and cans. Thirty new waste separation bins were deployed across the building and all old bins for general waste were removed from offices and com-

Timestamp	General 1	General 2	General 3	Paper 1
9-22-2014 8:30:00	70	0	0	35
9-23-2014 9:26:25	82	0	0	43
9-24-2014 9:40:00	83	79	71	66
9-25-2014 9:43:24	81	71	88	119
9-26-2014 9:56:18	88	76	90	60
Total (kg)	**144**	**70**	**99**	**68**

Table 6.7: Measurements before the waste separation process change

mon spaces. New bins were installed at edge locations of the building, within easy access of all users.

Timestamp	General 1	Paper 1	Paper 2	Plastic	Cans
10-2-2014 9:39	98	179	67	92	2
10-3-2014 9:39	109	169	115	66	2
10-6-2014 9:43	104	76	0	85	3
10-7-2014 9:50	97	113	79	70	2
10-8-2014 9:20	97	70	65	70	2
Total (kg)	**245**	**352**	**122**	**123**	**11**

Table 6.8: Measurements after the waste separation process change

Measurements were repeated by weighing the waste after the change in the waste separation process, as shown in Figure 6.8. The weight of waste was calculated by subtracting the weight of empty waste bins from the weight of the full waste bins. The new measurements took place in the beginning of October 2014. The results show that this change increased the amount of separated paper by 87% and reduced the amount of general waste by **22%**. It is also interesting to note that in only one week, 123 kg of plastic were recycled. Projecting this amount of plastic to the whole working year, the amount of plastic could increase to **5.535 kg/year**.

Even though our measurements were done manually, this process can be automated by having wireless scales connected to the GreenMind system, and presented to users as proposed in (Idsardi 2014).

Waste	Week total (before) [kg]	Week total (after) [kg]	Difference (%)
General waste	313	245	-21,73
Paper	68	474	+87,34
Plastic	0	123	+123
Cans	0	11	+11

Table 6.9: Comparison before and after the waste separation process change

6.2   Economic Considerations

Economic sustainability requires that a business or country uses its resources efficiently and responsibly so that it can operate in a sustainable manner to consistently produce an operational profit.[5] Many organizations and businesses are being approached by service companies promising interventions that may lead to reduction of their energy, water and other costs. The question arises how to evaluate different offered interventions and make an informed and calculated decisions.

One way to evaluate different possible interventions is by using economic evaluation. The parameters that investors most frequently take into account are the *Return on investment (ROI)* and the *Payback period*.

"Return on investment is the benefit to the investor resulting from an investment of some resource. A high ROI means the investment gains compare favorably to investment cost. In business, the purpose of the ROI metric is to measure, per period, rates of return on money invested in an economic entity in order to decide whether or not to undertake an investment. It is also used as indicator to compare different project investments within a project portfolio."[6]

The formula for the Return on investment is the following:

$$ReturnOnInvestment = \frac{(GainFromInvestment) - (CostOfInvestment)}{CostOfInvestment}$$

The return on investment is expressed as a percentage. "Return on investment may be calculated in terms other than financial gain. For example, social return on investment (SROI) is a principles-based method for measuring extra-financial value (i.e., environmental and social value not currently reflected in conventional financial accounts) relative to resources invested."[5]

Another parameter that serves to compare different solutions is the Payback period. Payback period means the period of time that a project (or intervention)

---

[5]http://www.circularecology.com/sustainability-and-sustainable-development.html
[6]https://en.wikipedia.org/wiki/Return_on_investment

requires to recover the money invested in it.[7] The payback period of a project is expressed in years and is computed using the following formula:

$$PaybackPeriod = \frac{CostOfInvestment}{NetAnnualCashInflow}$$

In the following section, to evaluate each saving solution performed within the Sustainable Bernoulliborg project, we will calculate and present: cost of investment, estimated or measured gain from investment, as well as return on investment and payback period. The return on investment is calculated for 10 years, being the customer requirement described in 1.5. Moreover, we assume that the average lifetime of each affected system will at least be 10 years. Thereafter, we discuss the factors that affect ROI and payback period and thus the economic acceptability of the project. One may notice that our primary focus is on the payback period as that is the parameter most often emphasized in discussions with the stakeholders. The ROI is also calculated and presented for business readers who may be more accustomed to use this parameter for comparison of solutions or projects.

6.2.1 Lighting Control

To evaluate the economic savings and payback period of our Lighting control solution, we first present the costs of investment for this solution.

	Pieces	Price per piece [EUR]	Cost [EUR]
Actuators	30	48.67	1460
Sensors	15	77	1185
Installation			0
Total:			**2645**

Table 6.10: Cost of investment for the Lighting control solution

Table 6.10 presents the cost of investment. For this intervention, the cost of investment is the sum of costs of purchased sensors and actuators. The software development costs and the installation costs were not taken into account as they were performed for the whole project by our team internally, and this work was not divided per implemented intervention. Furthermore, software development costs are considered to be part of R&D costs, and are considered to be a one-time investment when the system is being developed. Moreover, in this calculation we did not include power network or sensor housing preparation, as

---

[7]http://www.accountingformanagement.org/payback-method/

these tasks were performed by the *Facility Services department (Building and Land Management)* of the University of Groningen, responsible for the infrastructure.

Taking into account working/occupancy periods, number of controlled light fixtures and their average consumption, as well as measured savings of the Lighting control subsystem, we calculate the annual economic savings to be **338 EUR/year**, as presented in Table 6.11. It is important to note that the restaurant has a longer working/occupancy period than regular offices. While offices are occupied on an average of 45 weeks, the restaurant has only one non-working week, making the average number of working weeks to be 51.[8]

Electricity price and economic savings	
Working days	5
Working weeks	51
Number of light fixtures	30
Average time light fixtures are ON	13,50
Average power consumption per light fixture [W]	34,66
Annual energy consumption [kWh]	3580
Electricity price for University [EUR/kWh]	0,11
Annual electricity price for restaurant area* [EUR]	**393,75**
Measured savings by using SmartLighting system [%]	86
Annual economic saving [EUR]	**338,62**

Table 6.11: Economic savings of the Lighting control solution

The Payback period is calculated as the total cost of investment divided by the total annual savings (see Table 6.12). The return on the investment is calculated taking the period of 10 years as referent. Therefore, the payback period for this intervention is **7,81 years**, while the ROI is 28%.

Investment [EUR]	Annual saving [EUR]	Payback period [years]	ROI
2645	338	7,81	28%

Table 6.12: Payback period of the Lighting control solution

For Lighting control we foresee that the cost of investment can be reduced if the price of hardware (e.g., sensors and actuators), the number of hardware items used, installation and other factors are reduced. Moreover, the cost of hardware can be reduced by grouping more light fixtures to one actuator (e.g., one group per area), as well as using virtual (i.e., calculated) sensor values instead of installing more sensors.

---

[8]Personal communication with the restaurant manager.

For the Lighting control solution, the efficiency of the algorithm highly depends upon the type of space where the system is installed (e.g., a dark space without a lot of windows or a bright space with many windows), as well as upon the way the lights in the space were used before the intervention (e.g., whether the lights are scheduled to work 12 hours in a row or presence sensors have already been installed). Moreover, weather also plays an important role, as savings may vary depending on the weather conditions (e.g., being sunny or cloudy for more days a year). As previously mentioned, the Bernoulliborg restaurant is surrounded by windows which allow for significant harvesting of natural light. Moreover, in the situation before the intervention the lights were controlled in a scheduled way (the lights work 13,5 hours in a row). Both facts leave significant space for optimization by using location-based automated Lighting control.

6.2.2   Computer Sleep Mode Control

In this section we examine the economic savings that are achieved by deploying Computer Sleep Mode control to the PCs in the Bernoulliborg.

Item	Number of PCs	Cost per PC	Cost [EUR]
Installation	x	0	0

Table 6.13: Cost of investment for computer sleep mode control solution

In our project, there were practically no installation costs involved, as shown in the Table 6.13. However, in deployments to follow, there will always be costs related to installation.

That is why it is important to understand that for interventions involving a fixed average savings per item (PC) and fixed cost of investment per item, the payback period will be also fixed. In other words, if the installation costs are the only costs of investment, e.g., because there are no costs for hardware, and the average savings per PC are the same (in average y kWh/year), no matter whether the solution is installed in 14 or 300 PCs the duration of payback period will be the same. However, the total amount of savings will be higher in the case of more PCs affected. This flexibility enables adjustment of the payback period to a duration that is more acceptable to the customers.

Based on the experiment done by Setz (2015), the Computer Sleep Mode control solution will bring an average savings of 12,77 kWh/PC/year as shown in 6.14. This calculation was based on the assumption that a PC is effectively used for 45 weeks, taking into account vacation days and holidays for university employees in The Netherlands. In the case where the solution is installed in 14

Electricity price and economic savings	
Working days	5
Working weeks	45
Number of PCs	14
Number of PCs in the Bernoulliborg	300
Average PCs working time	5,00
Average power consumption per PC [W]	120,00
Annual energy consumption [kWh]	1890
Electricity price for University [EUR/kWh]	0,11
Annual electricity cost for PCs [EUR]	**207,90**
Annual saving per PC [kWh]	12,77
Annual kWh savings for PCs under experiment [kWh]	179
Projected annual economic saving for PCs under experiment [EUR]	**19,67**
Projected annual economic saving for all PCs in Bernoulliborg [EUR]	**421,44**

Table 6.14: Economic savings of computer sleep mode control solution

PCs, total annual economic savings is not financially significant, and amounts to 19,67 EUR. However, if the solution were installed in all 300 PCs present in the building, savings would be more than 21 times larger, amounting to **421 EUR/year**. Taking into account that no real installation costs are involved (e.g., installation could be done by automated script prepared by the internal team), these would be clear savings, amounting to 4.214 EUR within 10 years. If the users of the PCs would be to be more wasteful or the saving algorithm becomes more efficient, these savings could be higher.

Investment [EUR]	Annual saving [EUR]	Payback period [years]
0	19,67	0

Table 6.15: Payback period of computer sleep mode control solution

Finally, as presented in Figure 6.15 for this intervention the payback period is **0 years**, as the cost of investment is 0 EUR. For example, if there was a fixed installation cost of 10 EUR/PC, the payback period would be 7,12 years.

Even though the total savings achieved within the project are not economically significant, this solution provides valuable additional information about PC usage in a workspace. Using this information we can introduce additional savings for other control solutions by increasing the sensor data accuracy. For instance, PC usage information provides higher accuracy of algorithms for presence and activity recognition that are used for lighting, appliances and heating control.

6.2.3 Sensor Holder

The intermediate sensor holders were designed and developed in cooperation with a local manufacturing company.[9] The total price of production included design and development hours of manufacturers, as well as hardware
parts. Installation was also carried out by an external company and had related
costs. The total cost of investment for the sensor holders was 7.109 EUR, Table 6.16.

Item	Pieces	Price per piece [EUR]	Price [EUR]
Production	250	23	5.750
Installation	250	5,44	1.359
Total			7.109

Table 6.16: Cost of investment for sensor holder solution

As the original sensors work in a closed loop with lights, they do not have
any communication modules and do not store occupants' presence data. For
this reason we were not able to measure baseline data on average presence of
employees in offices. That is why we had to conduct a baseline survey (presented
in Appendix 1) to collect data from users and determine the economic savings
of this intervention. The survey was conducted among staff members working
in the Bernoulliborg. In total, 100 staff members filled in the survey. In this
survey we asked them about the average time they spent inside and outside their
offices during working days to determine average time when lights are turned
on as well as to estimate savings after the intervention, as presented in Table 6.17.
Projected annual savings for this intervention are **2.058 EUR/year**.

The payback period of this solution amounts to 3,45 years, while the ROI
is 190%, as shown in Table 6.18. As the sensor holders are installed in almost
all offices in the Bernoulliborg where such change could bring improvement,
total annual savings can be increased only when the offices are used in such a
way as to increase the efficiency of the solution. An example is when employees
frequently leave their offices for a period of 45 minutes to one hour (e.g., having
more meetings) and then come back to use it again. Savings in such cases would
increase, as the new shorter timeout would ensure that lights went off 15 minutes
after employees leave their offices instead of the previous 45 minutes.

The payback period could be shorter for every new installation, as design

[9]http://www.rug.nl/umcg/diensten/instrumentmakerij

Economic savings	
Working days	5
Working weeks	45
Number of offices	180
Average number of light fixtures per office	5,00
Average power consumption per light fixture [W]	44,00
Average time when lights are turned on [h]	7,41
Annual energy consumption [kWh]	66.023
Electricity price for University [EUR/kWh]	0,11
Annual electricity costs for office lights [EUR]	**7.262,54**
Estimated savings by using the system [%]	28,34
Projected annual economic saving [EUR]	**2.058,20**

Table 6.17: Economic savings of sensor holder solution

Investment [EUR]	Annual saving [EUR]	Payback period [years]	ROI
7.109,00	2.058,20	3,45	190%

Table 6.18: Payback period and ROI of sensor holder solution

costs would only be included the first time and the manufacturing process would become more efficient. Moreover, with an increased number of orders, the raw materials would also become less expensive.

Finally, one should note that baseline data was theoretically obtained from the users. As part of future work it would be interesting to repeat this study and recalculate savings based on actual sensor readings in offices. With our system, such as the Lighting Control system installed in the restaurant of the Bernoulli-borg, this is already possible.

6.2.4 Water Consumption Reduction

For the water consumption reduction intervention the cost of investment consisted of hardware devices (i.e., water flow reducers) and installation of the devices, as presented in Table 6.19. Moreover, additional water meters were installed so that savings could be measured. These costs are not included in the calculation as they are not a part of the intervention but represent a part of the evaluation process.

The average water consumption is 3.021 m^3. The university pays twice for water, once when the water is used and again when the used water is transported through the drain. As the price of m^3 of water (2,25 EUR) is higher than the

Item	Pieces	Price [EUR]	Cost [EUR]
Devices	35	5	175
Installation	35	14,29	500
Total:			**675**

Table 6.19: Cost of investment for water consumption reduction solution

	Economic savings	
	Consumption [m^3]	3.021
	Price [EUR/m^3]	**2,25**
Avg cost of annual water consumption [EUR]		6.797,25
Annual water savings of 5% [EUR]		**339,86**

Table 6.20: Economic savings of water consumption reduction solution

price of kWh of electricity (0,11 EUR), finding efficient ways to reduce water consumption may lead to higher economic savings. As presented in Table 6.20, annual water savings of 5% lead to savings of **339 EUR/year**.

Investment [EUR]	Annual saving [EUR]	Payback period [years]	ROI
675	339	**1,99**	404%

Table 6.21: Payback period and ROI of water consumption reduction solution

Table 6.21 shows that the payback period for this intervention amounts to **1,99 years**, with a ROI of 404%.

6.2.5 Waste Separation Process Change

The costs of investment for the waste separation process change are presented in Table 6.22. Due to the budget limitations of the project, the bins were purchased outside the Netherlands. For that reason, the cost of investment in the waste separation bins, besides the investment in the hardware (bins) itself also included import taxes and transportation. Moreover, it included additional rubber protection material that had to be installed on the bins to prevent people from hurting themselves while using the system. Installation costs were not calculated, as the installation was done by our internal team and did not involve direct costs.

According to official numbers gathered from the building manager, for 2012 the building management paid 11.312 EUR for waste management and transportation. From the results of the experiment explained earlier in the Sec-

Item	Pieces	Price per piece [EUR]	Price [EUR]
Waste separation bins	30	280,83	8.425
Import taxes	30	62,07	1.862
Transportation	30	28,83	865
Edge protection rubber	30	9,90	297
Installation	30	0	0
Total:		**381,63**	**11.449**

Table 6.22: Cost of investment for waste separation solution

Economic savings	
Annual costs general waste [EUR]	11.312
Measured savings using the system [%]	21,73
Savings on general waste reduction [EUR]	**2.458,09**

Table 6.23: Economic savings of waste separation solution

Investment [EUR]	Annual saving [EUR]	Payback period [years]	ROI
11.449,00	2.458,10	4,66	115%

Table 6.24: Payback period and ROI of waste separation solution

tion 6.1.5, we concluded that due to the increased recycling of paper, plastic and cans, the amount of general waste was reduced by 21,73%. For the whole building, that led to waste management savings of **2.458 EUR/year**.

According to the presented calculations, the payback period of this intervention is **4,66 years**. The payback period could be reduced if the University had a contract whereby additional revenues could be collected for recycled materials (paper, plastic, cans). Moreover, promotional campaigns could lead to an increase of recycling rates and reduction of amounts of general waste.

6.2.6 Research, Development and Project Costs

The remaining costs, not represented in previous sections, were costs related to research and development. As the research and development of the mentioned solutions were done as the part of the Sustainable Bernoulliborg project, the remaining costs covered the time spent by researchers, assistant researchers, software developers, etc. Moreover, other project costs such as promotional materials, meeting costs, experimental equipment, evaluation and experiment preparation costs, and general management costs were included in the total cost of the project. If these costs were taken into account, the payback period would increase and the economic acceptance would decrease.

6.2.7 Carbon Footprint

In reading this thesis, one may have expected to see how much the carbon footprint (quantified as amount of emitted greenhouse gases, carbon dioxide equivalents) is reduced by each solution. However, we deliberately chose not to calculate reduction of the carbon footprint by use of the described interventions. Our main reason is that the carbon footprint is very hard to calculate precisely (Berners-Lee 2010). To define the total carbon footprint one should take into account both direct and indirect emissions. For example, direct emissions that are reduced by optimizing lighting use can easily be canceled by indirect emissions caused by using long transportation of the equipment needed to implement the intervention. Tracing back all the things that have to happen to make the interventions leads to an infinite number of pathways (Berners-Lee 2010). In the end, one would need to calculate the emissions caused by writing and printing this thesis, as well as the number of coffees one has drunk during the writing, electricity consumed by the laptop while writing the thesis, etc. However, even though we have in this thesis not calculated the reduced carbon footprint, we have presented actions that can be taken to reduce it.

Chapter 7

Social Acceptance Considerations

Computer systems can improve organizational performance only if they are used. Unfortunately, resistance to end-user systems by managers and professionals is a widespread problem (Davis et al. 1989).

To evaluate the social acceptability of the GreenMind system, we use the constructs of the Technology Acceptance Model - TAM (Davis 1986). According to TAM, a user will have a positive attitude toward using a (computer) system if he or she perceives that system as easy to use as well as useful, and a positive attitude leads to actual system use. The TAM has been continuously studied and expanded in the two major upgrades: the TAM 2 (Venkatesh and Davis 2000), (Venkatesh 2000) and the Unified Theory of Acceptance and Use of Technology or UTAUT (Venkatesh et al. 2003).

Users include both energy/facility managers, who use more advanced features of the system to provide additional optimizations of building use, and end-users (i.e., building occupants or visitors), who use basic system functionalities necessary for their work.

To measure *Perceived Usefulness* and *Perceived Ease of Use* of the GreenMind system by the energy and facility managers (i.e., persons who have crucial roles in the decision making process about adoption of the system) we interviewed the subjects – previously identified as stakeholders – to evaluate their intention to use the system. These inputs were gathered through a survey, presented in Appendix 2. Moreover, as the GreenMind system has already been implemented, and as end-users are already using the building, we also evaluated their satisfaction and/or acceptability of the system as based on actual system use. It is important to mention that some parts of the system involved voluntary use (e.g., Computer Sleep Mode control), while others involved mandatory use (e.g., Lighting control). The end-user opinion coming from actual system use was measured by resorting to several surveys, containing questions about the usability, satisfaction or acceptability of the system, where applicable.

Below we discuss social acceptability per affected system or intervention. In the Conclusions chapter (Chapter 9) we present the social acceptability of the overall system. During preparation of the surveys, colleagues from the University of Groningen, the Faculty of Behavioural and Social Sciences, and Environmental Psychology Department of Social Psychology (see Acknowledgments) provided valuable input as well as revision of some of the surveys.

7.1 Consumption Display

The function of the *Consumption display* component is to present consumption data with a goal to inform and motivate users to conserve resources (i.e., energy, water). This component is not responsible for any control of energy consuming devices and therefore cannot be evaluated with respect to environmental savings and economic acceptability. However, it can be evaluated with respect to its social acceptability.

The main goals of the Consumption display are to provide building occupants with information to raise their awareness of energy consumption in the building, and to motivate them to perform more frequent energy saving actions. In order to evaluate whether user awareness was raised and whether that led to positive actions, we designed a survey asking building occupants to state their level of agreement regarding increased awareness, motivation to perform saving actions at work as well as at home, the acceptability of the intervention and finally satisfaction with the quality of the solution.

The survey was sent to a mailing list containing all building occupants of the Bernoulliborg and was filled in by 63 occupants. The complete survey can be found in the Appendix 6, and its results can be seen in the Figure 7.1.

The figure shows that 41,27% of survey participants experienced increased awareness, while 34,92% participants did not. More than one-fifth of the participants gave a neutral answer. The information provided by the consumption display motivated only 9,53% participants to perform saving actions more frequently at work, while it did not motivate the majority (55,56%) of participants. 31,75% of participants gave a neutral answer. Surprisingly, 22,22% of the participants answered that they more frequently performed energy saving actions at home after being motivated by the power consumption display, but a majority of 53,97 were not so motivated. 42,86% of the participants were satisfied with the quality of the consumption display, 22,22% thought that it could be improved, and 28,57% were neutral.

Finally, the acceptability of this intervention was rather high, with a total of **66,66%** participants giving a positive mark (20,63% strongly agree and 46,03% agree). Furthermore, invaluable feedback for improvement of the consumption display was collected and will be used in the process of further development.

These responses show that our intervention raised awareness for a large

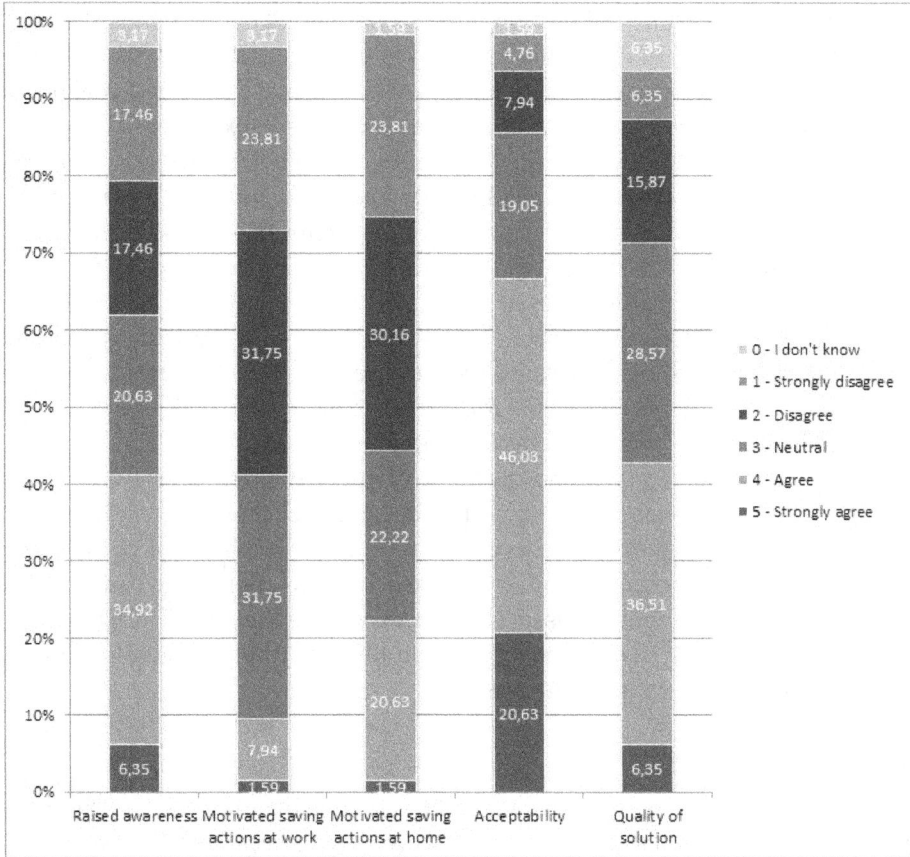

Figure 7.1: The results of the social acceptability study for the consumption display intervention

number of surveyed participants and that they consider it to be acceptable. It would be interesting to measure the actual energy and economic savings caused by providing feedback to building users. However, for such an experiment the intervention should be done in an isolated way, with no other interventions to affect the results. Moreover, certain methods should also be applied to rule out the effects of weather (e.g, degree-days methodology).

7.2 Lighting Control

In this section, we evaluate the Lighting control solution from a social acceptability perspective. To do so we prepared a questionnaire to test the solution for usability. We adopt the usability definition from the ISO 9241 standard for ergonomics of human-computer interaction that states: "usability is the extent to which a product can be used by specified users to achieve specified goals with effectiveness, efficiency and satisfaction in a specified context of use".

We evaluated the *acceptability*, learnability, system effectiveness and efficiency, as well as usefulness of the Lighting control system. Acceptability indicates the attitude of end-users towards our system. This includes the use of sensors, switching on of lamps, etc. Learnability refers to the need for users to understand how to use our system (e.g., do users know how to trigger the lamps). System effectiveness refers to the satisfaction of users with the overall system. Efficiency refers to the satisfaction of users with the time needed for the system to perform its tasks. While this aspect is more technical, users can still evaluate how they perceive the effectiveness of our system.

Since the system is integrated unobtrusively into the environment, its use is not actual or intentional. Most users do not even notice that a system to control lighting has been installed. However, users do expect an amount of light that will enable them to perform their actions without distractions, e.g., eating lunch or reading a book. Moreover, users may not expect the same level of light while eating lunch as when reading or working in the restaurant outside lunchtime. This survey can be seen in the Appendix 4.

The Survey Participants

We describe the survey participants and their division into groups. The questionnaire was conducted for two types of users, one experiencing the system during lunchtime (*Lunchtime Group*), and the other outside lunchtime (*Outside Lunchtime Group*).

Inputs for the *Lunchtime Group* were collected on the 7th and 9th of April, 2015. The total number of inputs was 54. Most subjects were visitors at the restaurant (57,41%) while others were occupants who work (25,93%) or study (16,67%) in the building. Most subjects used the restaurant only for lunch (96,30%) while the rest used it both for lunch as well as study or work. The *Lunchtime Group* consisted of subjects who consider themselves to be aware of sustainability issues (83.33%) and those who engage in environmentally friendly behaviour (79,63%). A high percentage of the subjects were familiar with automated control in buildings (64,81%) and lamps triggered by movement sensors (83,33%).

Inputs for the *Outside Lunchtime Group* were collected on the 4th and 7th of May, 2015. The total number of inputs was 18. Most of these subjects were students (72,22%) and visitors (27,78%) who use the restaurant both for studying/-

working and eating lunch (61,11%), some only for eating lunch (22,22), 11,11% for playing games and 5,55% for doing business. The *Outside Lunchtime Group* also consisted of subjects who consider themselves aware of sustainability issues (77,78%) and those who engage in environmentally friendly behaviour (88,89%). More than half of these subjects were familiar with automated control in buildings (55.55%) and lamps triggered by movement sensors (66,67%).

The Survey Results

The results of the questionnaire were divided per: awareness, perception, learnability, acceptability, efficiency, effectiveness, and usefulness. Each aspect is represented by one or more items in the questionnaire. We discuss below the results for each aspect per questioned group.

As shown in Figure 7.2, the *Lunchtime Group* showed only partial *awareness*, with one third of the subjects stating that they were aware of the system. 25,92% stating that they were not aware of the system, while 25,93% answered "I dont know. The lack of some subjects awareness of the system is actually good, as it proves that the system was unobtrusive, enabling energy saving without affecting user comfort. The *perception* of the *Lunchtime Group* was that this system saves energy (64,81%), takes into account the natural light level (53,70%) and considers people's presence (72,22%). This implies that the majority of this group had good assumptions about the system or were simply well informed.

Regarding *learnability*, 42,59% of the group stated that it was easy to use the system, while 51,85% were neutral or did not know the answer to this question. 71,11% of the subjects found the system to be *acceptable*, while 16,66% thought it caused distractions.

As for the *efficiency* of the system, the majority of subjects did not know whether the system immediately reacted to change, or gave a neutral answer (61,11%). 24,07% considered the system to be efficient, while 14,81% did not. When it comes to *effectiveness*, 59,26% of the subjects stated that they were satisfied with the system, 31,48% were neutral and only 5,56% were dissatisfied, as shown in Figure 7.3. Finally, the majority of subjects (70,37%) stated that they found the system to be *useful*, with 12,96% neutral and only 11.14% regarding it as not useful.

The *Outside Lunchtime Group* showed moderate *awareness* with a bit more than one third of the subjects (38,89%) stating that they were aware of the system. 25,92% of subjects stated that they were not aware of the system, while 25,93% did not know the answer to this question (Figure 7.4). When compared with the *Lunchtime Group*, this group had a slightly higher awareness of the system. This may be because this group usually worked outside lunch hours, sometimes in

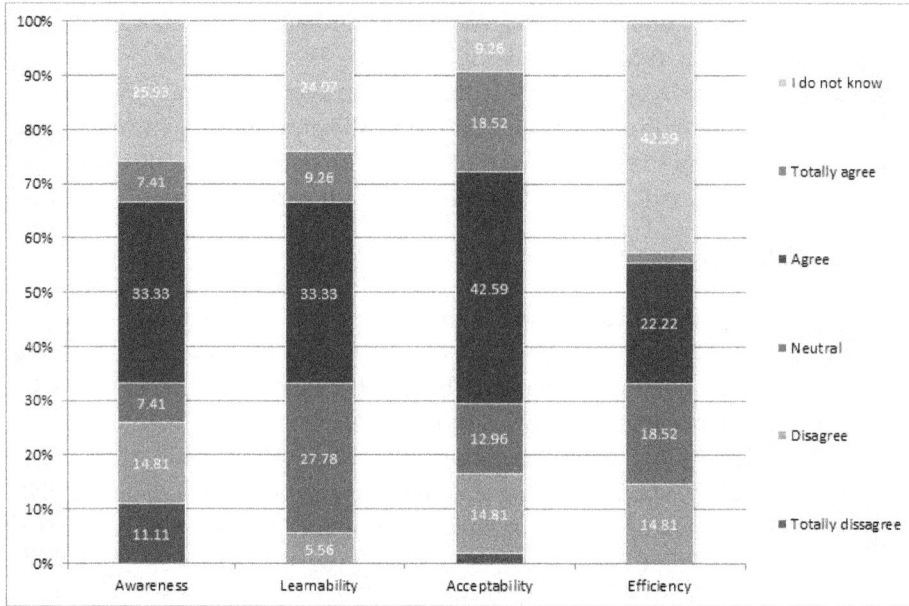

Figure 7.2: Awareness, Learnability, Acceptability and Efficiency for the *Lunchtime Group*

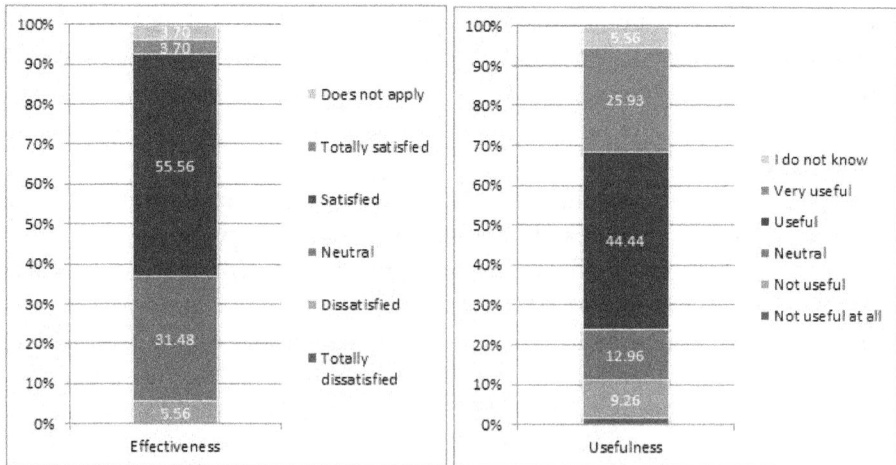

Figure 7.3: Effectiveness and usefulness for the *Lunchtime Group*

the late afternoon or evening, when changes in the system became more notice-able.

The *perception* of the *Outside Lunchtime Group* was that this system saves energy (77,78%), and considers people's presence (66,66%). Only half of the sub-jects thought that the system considers the natural light level. This may be be-cause the system deployed in their offices takes only presence/movement into account, and end-users assumed that the same system was used in the restau-rant. However, we may conclude that the majority of this group were well in-formed regarding energy saving and presence detection. Moreover, to better inform end-users and thus increase their acceptance of the system, they need better communication about the systems inclusion of the natural light level. Re-garding *learnability*, 66,67% of the group stated that it was easy to use the system, while 33,33% were neutral or did not know the answer to this question. 83,33% of the subjects found the system to be *acceptable*, while none of the subjects con-sidered the system to cause distractions. This is a very good result, as it confirms that this solution achieves the goal of saving energy without affecting user com-fort or productivity. As for the *efficiency* of the system, the majority of subjects did not know if the system immediately reacted to changes, or gave a neutral answer (72,22%). 22,23% found that the system was efficient, while 5,56% did not agree. Figure 7.5, shows that when it comes to *effectiveness*, 72,23% of the subjects stated that they were satisfied or very satisfied with the system, 11,11% were neutral and none were dissatisfied. Most importantly, the majority of sub-jects (83,33%) stated that they found the system to be *useful*, with 5,56% neutral and only 11,11% stating that they did not know the answer.

Referring back to the constructs of the technology acceptance model, we may conclude that both groups showed high acceptance of the system (Lunchtime Group **71,11%** and *Outside Lunchtime Group* **83,33%**) and high perceived useful-ness (Lunchtime Group 70,37% and *Outside Lunchtime Group* 83,33%). We take average value from both surveyed groups to be the final result regarding accept-ability of this study (**77.22%**). The perceived ease of use is equivalent to learnabil-ity, which showed to have slightly lower yet satisfactory results (*Lunchtime Group* 42,59% and *Outside Lunchtime Group* 66,67%). All in all, the system can be con-sidered to be well received by the end-users, and since there were no complaints during the whole period when the system was used, we can easily recommend deploying this system to other locations.

7.3   Computer Sleep Mode Control

Even though the Computer Sleep Mode control solution was deployed on fourteen PCs, to evaluate social acceptance only five end-users responded to the survey. The survey consisted of six questions and is presented in Appendix 5.

When asked if they turn off their computer before leaving the office, four out

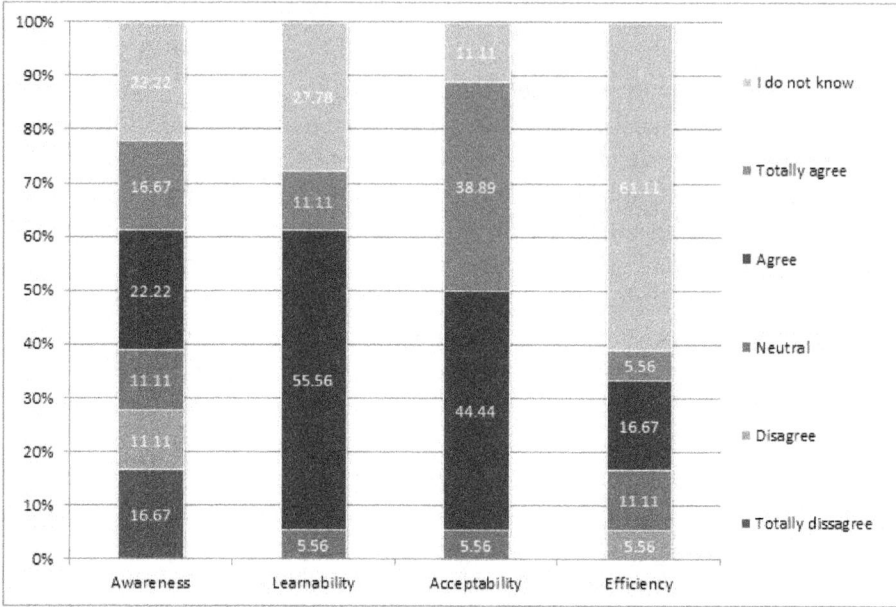

Figure 7.4: Awareness, Learnability, Acceptability and Efficiency for the *Outside Lunchtime Group*

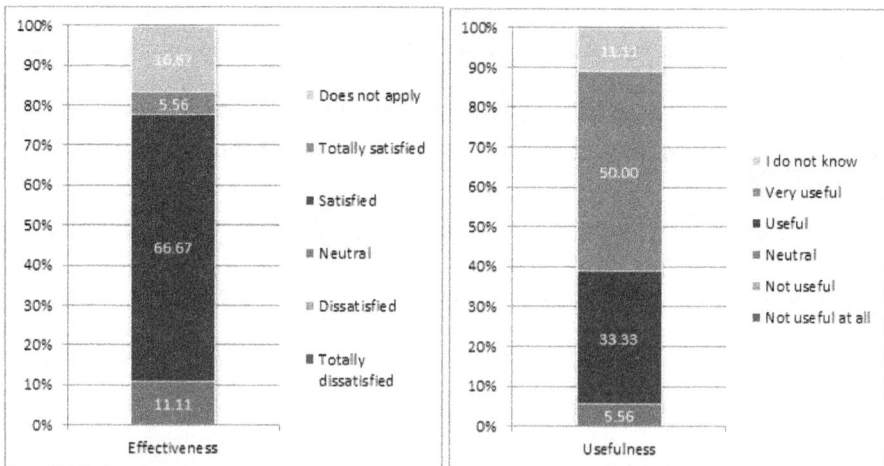

Figure 7.5: Effectiveness and usefulness for the *Outside Lunchtime Group*

of five replied "Always" and one replied "Frequently". This implies that these users are already quite environmentally aware, in the habit of turning off their workstation at the end of the working day, or both.

When asked whether the enabling of the sleep mode made them more aware of how they use their computers, three out of five agreed, one was neutral and one disagreed.

The end-users were also asked whether enabling of the sleep mode made them put the computer to sleep more often. Four out of five answers were "Agree" (one Strongly Agree), while one was neutral. Also, for the question whether enabling of the sleep mode made them turn the computer off more often, two out of five agreed, while three disagreed.

To determine the frequency of their disturbance by the solution, we asked the end-users if on some occasions the computer entered sleep mode while they were actively using the computer. Two out of five answers were "Never", one "Rarely", one "Occasionally" and one "Frequently". This showed rather different usage experience and frequency of disturbance.

Finally, when asked if having to wake up their computer from sleep mode disrupts their work flow, three out of five surveyed end-users stated "Not much", while two stated "Little".

Finally, as there were too few sufficient survey responses by end-users to draw statistically relevant conclusions, we were not able to completely evaluate this solution from the perspective of user acceptability.

7.4 Sensor Holder

The intermediate sensor holder intervention was also evaluated from the social acceptability perspective. The survey was sent to the mailing list of all building occupants. The participants were asked to mark their level of agreement to the following three statements: (1) Adjustment of existing movement sensor increased my satisfaction with how the lights are controlled within my office, (2) I was satisfied with the quality of the added sensor holder, and (3) In my opinion, existing movement sensor adjustment was an acceptable intervention.

In total, 63 building occupants filled in the survey (see Appendix 6). The Figure 7.6 shows the results of this evaluation study. The results show that there was an increase in satisfaction for 39,69% survey participants. However, there were also 28,57% participants who stated that this solution gave no increase in satisfaction. Results regarding the quality of the solution were very similar. Again, a majority of 42,86% survey participants stated that they were satisfied with the quality of the solution, while 25,39% were not. For this and previous statements, 19,05% participants answered that they did not know. Finally, **55,56%** survey participants found this intervention to be acceptable; we therefore marked it with "medium" acceptability. Possible reasons for medium acceptability may

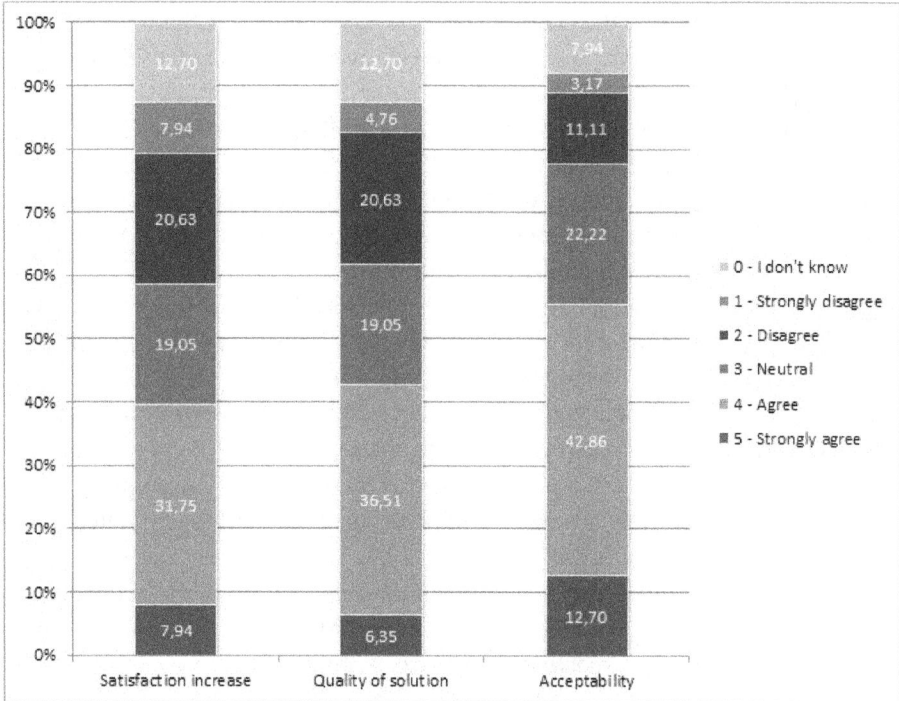

Figure 7.6: The results of the social acceptability study for the intermediate sensor holder intervention

be that some users already had well-positioned sensors and our change did not demonstrate clear improvements. Moreover, similar reasons may apply to users who are very active and move a lot during their work day. For them the benefits of this intervention are not sufficiently visible.

7.5   Water Consumption Reduction

We also used a survey to evaluate the water consumption reduction intervention. In this survey we asked the building occupants to state their level of agreement on the following four statements: (1) Adding water flow reductors to water faucets in the building did not decrease my satisfaction regarding water usage from faucets; (2) Motivated by water flow reduction action, I performed water saving actions at work more frequently; (3) Motivated by the water flow reduction action, I performed water saving actions at home more frequently; (4) The water flow reduction was an acceptable intervention.

In total, 63 building occupants filled in the survey. Figure 7.7 shows the results of this study.

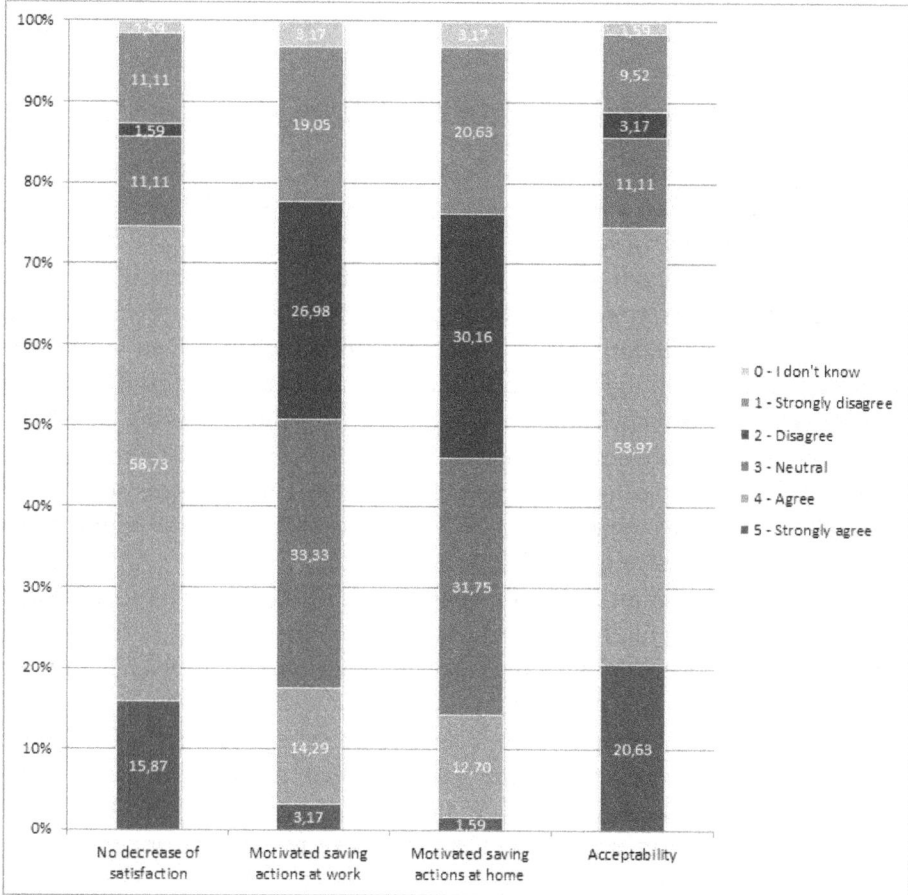

Figure 7.7: The results of the social acceptability study for the water consumption reduction intervention

Survey results show that 74,6% of the participants did not experience a decrease in satisfaction with this intervention, while 12,7% of survey participants did experience a decrease in satisfaction. From the comments in the survey we conclude that the dissatisfaction was experienced due to longer waiting time when using a water faucet (e.g., when filling a bottle).

The results also show that a small percentage of participants were motivated by this intervention (17,46%) to perform water saving actions at work. The results, however, imply that there was no significant spillover effect from the work to the home situation.

Finally, **74,6%** of survey participants stated that this intervention was acceptable. Taking into account this percentage, we mark this action as highly acceptable.

7.6   Waste Separation Process Change

As part of the waste separation intervention, all trash bins were removed from offices and replaced by new waste separation bins in the hallways. This allowed for separation of waste into paper, plastic, glass and cans.

The intervention was top-down, meaning that the change occurred over night and without input from people working in the Bernoulliborg. In the current study, we aimed at finding out how the intervention was received by the employees. More specifically, we aimed to investigate whether employees found the intervention acceptable, and whether it influenced their recycling behavior.

We also measured the acceptability of this intervention and recycling rates among employees working in a neighboring building of the same faculty and the same campus, the Linnaeusborg[1], where employees still use the regular trash bins; this group served as a control group.

To test the effectiveness of the intervention we carried out an online study and asked participants to fill in a short questionnaire. The questionnaire included scales measuring acceptability of the recycling policy, current recycling behavior at work and ease of participating in the intervention (effort ratings).

This study is a joint work with Ellen van der Werff and Berfu Unal from the Faculty of Behavioural and Social Sciences, Environmental Psychology Department Social Psychology, as well as Tuan Anh Nguyen from the Distributed Systems research group. Data collection was performed by Distributed System group members, while data analysis was performed by members of the Social Psychology department.

The Survey Results

A total of 266 employees participated in the online study. The majority of participants (140) were in the experimental situation, and therefore working in the Bernoulliborg. The rest of the participants (126) were in the control situation, working at the Linnaeusborg building.

Participants in the experimental situation read about the intervention, and participants in the control condition read about a hypothetical version of the

---

[1]http://www.rug.nl/about-us/who-are-we/discover-groningen/linnaeusborg

same intervention (e.g., "Imagine that the Faculty Board decided to remove regular trash bins, and replace them with new ones that require separating your trash."). Then they answered the same questions about the general acceptability of this policy. The answers are given on a scale of 1 to 7, and the results are presented as average value for all responses. Results revealed that acceptability was significantly higher in the experimental situation (5,56 out of 7; **79,43%**) than in the control situation (4,99 out of 7; 71,29%). This finding indicates that participating in the intervention already increases acceptance of the policy, a finding that is in line with previous research.

Participants were asked to indicate the extent to which they recycle and separate their trash at work. As expected, average self-reported recycling was significantly higher in the experimental situation (5,37) than in the control situation (4,14). The finding indicates that providing employees with new recycling bins already results in higher separation of trash.

Participants indicated to what extent they aim to reduce waste at work. We found that those in the control situation (4,88) were more likely to reduce waste at work than those in the experimental situation (4,28). This may suggest that there is more willingness to reduce waste at work in cases where comfort was not affected because the intervention did not actually take place.

Participants were asked to indicate the extent to which it is easy for them to recycle trash at work. There were no differences between the experimental and control groups in effort ratings (4,99; 5,25) respectively. This finding indicates that employees who were already participating in the intervention by separating their trash in the Bernoulliborg did not find this more taxing or difficult than did employees who were not yet actively recycling their trash at work.

The results of this study show that introducing this policy will increase recycling rates in university buildings. Furthermore, people who experience this policy in their own building find it more acceptable than do people who do not yet experience it. Overall, the participants rated the policy as rather acceptable. These results suggest that experiencing the policy in one's own building even increases acceptability of the policy.

However, our findings also suggest that introducing such a policy may have some negative consequences. We found that participants in the building with the new recycling bins were less likely to reduce their waste than participants in the building without recycling bins. This suggests that the intervention may be successful in promoting recycling, but at the same time reduce the need to reduce waste. An important question is therefore whether the policy can be adjusted in such a way as to promote not only recycling, but also other sustainable behaviors such as reducing waste. A possible approach worth exploring would be to adjust the policy by making it a more bottom-up approach in which the employees are involved at an earlier stage and in which the goal of the policy is clearly communicated. We assume that this will improve the effectiveness of the policy regarding other sustainable behavior such as reducing waste.

Chapter 8

Related work

The problem of energy efficiency is not new and it is tackled by researchers from different disciplines. Because of the multidisciplinary nature of this thesis, we present scientific research related to energy efficient buildings, from computer science, psychological and economic perspectives. To address the work related to design and implementation of the GreenMind system, we provide details on related software architectures, as well as application of related artificial intelligence algorithms. Subsequently, we present both already implemented and ongoing research projects, and define similarities and differences with the Sustainable Bernoulliborg project. Finally, we present several commercial products providing partial solutions to the problem of energy efficiency in buildings, as well as present commercial services offered today.

8.1 Scientific Research

In (Nguyen and Aiello 2013), the authors provide a survey about energy intelligent systems looking at the three main energy consuming subsystems in buildings (i.e., HVAC, lighting and office equipment) which have drawn the attention of numerous studies. The authors notice that some of the researched studies focus on one subsystem only, while others try to save energy for two or even three subsystems. The majority of the reviewed studies focus on only one subsystem. The same study shows that only six of all identified studies focus on HVAC and lighting, three focus on HVAC, lighting and power plugs, and two others focus on lighting and power plugs only. More specifically, smart lighting has been the focus of many authors (Dubois and Blomsterberg 2011, Bülow-Hübe 2008) as an area with significant energy improvement potentials. Besides lighting, office equipment (e.g., computer and monitor) influences the energy usage to a great extent (Kawamoto et al. 2004, Webber et al. 2006). This leads to phantom consumption or other unnecessary energy consumption as equipment stays turned on when there is absolutely no need for that. Furthermore, according to (Boyano et al. 2013, Agarwal et al. 2010), HVAC systems can also contribute to the possible energy and cost savings.

Contrary to this, in the present work, we focus on integration of lighting, workstations (PCs) and plug loads subsystems. Moreover, we provide easily extensible infrastructure that leaves space for integration of new subsystems, for example subsystems to control of heating, cooling or ventilation. By integrating the mentioned subsystems, we reuse the information and increase accuracy of presence and activity detection algorithms to further optimize energy use in non-residential buildings.

We also identified two interesting work on energy management in smart spaces. In (Caruso et al. 2014), the authors developed the OPlatform, for smart environments to be able to collect micro-account energy consumption of devices, at the level of each single power line, which allows at the same time the actuation of devices. A similar work deals with user profiling and micro-accounting for smart energy management (Caruso et al. 2013). These papers address similar issues that we tackle as a part of our User Layer solutions, described in Section 4.3.1.

There is number of studies that relate to psychological (social or behavioral) as well as economic aspects of energy efficiency in buildings. To identify the related work we performed a search using the terms "energy efficient buildings" with terms "psychological, social, behavioral" as well as with the term "economic".

A cross-cultural analysis of household energy use behavior in Japan and Norway (Wilhite et al. 1996) compares energy use behavior for space heating, lighting and hot water use, as well as discuss patterns related to cultural and economic factors.

The next identified study concerns efficient and inefficient aspects of residential energy behavior and what are the policy instruments for change (Lindén et al. 2006). In this study, the authors did an empirical study based on a survey of 600 Swedish households. The conducted interviews questioned about residential energy behavior and possible policy instruments for change.

Further, in (Darby et al. 2006) the authors explore feedback on household electricity consumption and if it represents a tool for saving energy. In their work, a psychological model is presented. Relevant features of feedback are identified that may determine its effectiveness: frequency, duration, content, breakdown, medium and way of presentation, comparisons, and combination with other instruments.

A study on energy conservation behavior and the difficult path from information to action is presented in (Costanzo et al. 1986). This study presents a social-psychological model of energy-use behavior that draws on behavioral and social research to explain influence processes and behavioral change related to

energy conservation behavior. Psychological factors that refer to how information is processed by individual decision makers and positional factors that relate
to characteristics of the decision makers' situations that support or constrain action were discussed.

Moreover, in (Poortinga et al. 2003) the authors present household preferences for energy-saving measures. In this paper, the authors conclude that:
"Energy-saving measures differed in the domain of energy savings, energy-
saving strategy, and the amount of energy savings . Energy-saving strategy
appeared to be the most important characteristic influencing the acceptability
of energy-saving measures. In general, technical improvements were preferred
over behavioral measures and especially shifts in consumption. Further, home
energy-saving measures were more acceptable than transport energy-saving
measures. The amount of energy savings was the least important characteristic: there was hardly any difference in the acceptability of measures with small
and large energy savings."

Furthermore, the factors influencing household energy use were studied
in (Steg 2008). In this work, the authors discuss three barriers to fossil fuel energy conservation, namely "insufficient knowledge of effective ways to reduce
household energy use, the low priority and high costs of energy savings, and the
lack of feasible alternatives."

Additionally, energy saving and energy efficiency concepts for policy making
were presented in (Oikonomou et al. 2009). In this paper, the authors attempt to
identify the effects of parameters that determine energy saving behavior with
the use of the microeconomic theory. The role of these parameters is crucial and
can determine the outcome of energy efficiency policies; therefore policymakers
should properly address them when designing policies.

As it can be seen, most of the above-mentioned psychological studies relate
to households and the residential sector. It would be very interesting to conduct similar studies with the focus on non-residential (especially office) buildings which, to the best of our knowledge, are currently unavailable.

Turning to the economic perspective, there are two studies in line with our
business scalability requirements, one on management of emerging technologies (Groen and Walsh 2013), and another on guidelines for creating a corporate
entrepreneurship function to realize business development in a high-tech context (Uittenbogaard et al. 2005). Furthermore, we identify a work that shows how
combining activism and entrepreneurship can contribute to online advocacy organizations promoting sustainability (drs. T.A. van den Broek et al. 2012). These
studies are helpful in as sense that they provide business perspective necessary
for technical teams to understand importance of business requirements and to

include those in every day development planning and execution.

In the work on delivering energy efficient buildings – a design procedure to demonstrate environmental and economic benefits (Horsley et al. 2003), the authors describe the development of a design management procedure in which energy performance is monitored from the earliest phases of building inception. Moreover, the decision support tool to give guidance to design teams at a stage in the design process on project-specific energy performance issues, and their environmental and economic implications was developed. The authors conclude that "Reducing the energy consumption of buildings represents not only a significant environmental improvement, but is also favorable over a medium- to long-term economic basis. Considering the likely rise in energy prices over the lives of new buildings, these economic benefits look set to become very significant indeed."

Next, we found a work regarding a methodology for economic efficient design of Net Zero Energy Buildings (Kapsalaki et al. 2012). In this work, the authors developed a methodology and an associated calculation platform in order to identify the economic efficient design solutions for residential Net Zero Energy Building (NZEB) design considering the influence of the local climate, the endogenous energy resources and the local economic conditions. It has been concluded that "as a general trend, the most expensive NZEB design solution in terms of initial cost was at least 3 times more expensive than the cheapest design solution. The same at least about 3:1 ratio, was generally observed in terms of life cycle cost. This result clearly illustrates the importance of an economic analysis at the early design stage of NZEBs in order to reach the energy goals with economic efficiency." As in the present work we are dealing with an existing building,this methodology could not be applied or taken into account.

In (Amstalden et al. 2007), the authors investigate economic potential of energy-efficient retrofitting in the Swiss residential building sector and the effects of policy instruments and energy price expectations. This work analyses the profitability of energy-efficient retrofit investments in the Swiss residential building sector from the house owner's perspective. Different energy price expectations, policy instruments such as subsidies, income tax deduction and a carbon tax, as well as potential future cost degression of energy efficiency measures were taken into account. The authors conclude that: "For the current economic assessment of energy-efficient retrofitting, two relevant factors were identified. First, the expected energy price has a significant influence on the outcome of the investment analysis. Second, the inclusion of financial energy policy support in the investment analysis is crucial. The implications for house owners and investors are also of interest. If energy prices remain high and current policy measures are taken into account, energy-saving retrofits would be highly attractive investment opportunities. Therefore, it would be economically wrong to renovate a house without simultaneous investment into energy efficiency."

The study on willingness to pay for energy-saving measures in residential buildings uses a choice experiment to evaluate the consumers' willingness to pay for energy-saving measures in Switzerland's residential buildings (Banfi et al. 2008). "The results show a significant willingness to pay for energy-efficiency attributes of rental apartments and of purchased houses. The willingness to pay varies between 3% of the price for an enhanced insulated facade (in comparison to a standard insulation) and 8% to 13% of the price for a ventilation system in new buildings or insulated windows in old buildings (compared to old windows) respectively."

The following paper explores economic returns to energy-efficient investments in the housing market, providing evidence from Singapore (Deng et al. 2012). Throughout this work, 250 building projects in the City of Singapore awarded the Green Mark for energy efficiency and sustainability were analyzed. More specifically, the private returns on these investments, evaluating the premium in asset values they command in the market were analyses. In total, almost 37,000 transactions in the Singapore housing market were analyzed to estimate the economic impact of the Green Mark program on Singapore's residential sector. The results show that based on nearest one-to-one neighbor matching between control and treatment samples, a significant premium in selling prices for dwellings with Green Mark Certification. The estimated premium is larger for dwellings certified at higher levels in the Green Market process Platinum, Gold Plus, and Gold rated dwellings. Comparing to the present work, we may draw a parallel and conclude that improving sustainability of a building does not bring only return on investment from energy savings, but can additionally increase the value of property by making it more attractive interested parties (e.g., tenants, buyers, etc.).

Finally, we present a paper on economic evaluation of energy saving measures in a common type of Greek building (Nikolaidis et al. 2009). This paper deals with the economic analysis and evaluation of various energy saving measures in the building sector, focusing on a domestic detached house in Greece, i.e. in a typical Mediterranean climate. In this paper, the authors conclude that "Using the Internal Rate of Return as evaluation criterion it has been shown that the upgrading of artificial lighting is the most effective investment, while the insulation as well as the installation of an automatic temperature control system at the burner boiler system follow next. The use of solar heaters is economic enough and profitable, contrary to the replacement of windows and door frames and the partial upgrading of heating systems that constitute very low return investments." This work confirms that besides retrofitting solutions, automated control of large energy consumers show to be effective investments.

Similar to presented economic studies, our present work implies that taking the system proposed in this work into account in early stages of design building owners or investors can save on operational costs and also reduce total cost of ownership if the investment is done at initially instead of retrofitting being done in later stages of building life-cycle. Moreover, in Section 6.2 we take economic considerations into account, namely return on investment and payback periods. Having constantly raising prices of energy in mind, we show how better control of energy consuming subsystems or retrofitting represent interesting investment opportunities. In present work (Chapter 6), we evaluated several investments as well as stakeholders willingness to invest if factors, such as payback period, become shorter.

8.2   Software Architectures and AI Algorithms

There are several initiatives aiming to make homes and offices, energy efficient and comfortable. In these initiatives, system architectures have been designed to satisfy common requirements. Starting from year 2000 onwards, there are a few projects which describe architectures of pervasive systems (see Table 8.1).

Inspired by work in (Degeler and Lazovik 2013), we looked further into the architectures of pervasive systems, presented in Figure 8.1. The presented architectures vary in organizational aspect (e.g., number of layers, components per layer), application scope (home, classroom, office(s), building) as well as level of implementation (from proof-of-concept, through living-lab until fully implemented projects in the real-world environment). In the following, we list the projects that are implemented and have their architecture described in a form of a publication.

There are several main differences between architectures described in the mentioned projects and the architecture that we propose. Most of the projects focus on individual homes or multiple offices, while our focus is a whole building and all building subsystems (e.g., Lighting, PCs, HVAC, appliances), therefore covering different use cases, scenarios and energy saving strategies. The listed projects cover less than 50 end-users (occupants), while some parts of our system are designed to cover more than 800 building users (e.g., the restaurant of the Bernoulliborg). For implementation each project uses different middleware implementation. We consider each component of the proposed architecture as a service, providing wrapper around actual devices, that way exposing function-

Project title	Started	Scope	Publication
MavHome	2000	Home	Youngblood, et al., 2004
iSpace/iDorm	2002	Dormitory room	Holmes, et al., 2002
SmartLab	2006	n/a	Lopez-de-Ipina, et al., 2008
CASAS SmartHome	2008	Duplex apartment	Kusznir & Cook, 2010
SmartHomes4All	2008	Apartment	Aiello, et al., 2011
ThinkHome	2010	Home	Reinisch, et al., 2010
EnergySmartOffices	2011	Offices	T. A. Nguyen, et al., 2013
GreenerBuildings	2013	Offices	Degeler et al., 2013
SB / GreenMind	2014	Office building	Nizamic, et al., 2014

Table 8.1: Projects introducing architectures in pervasive systems

alities and making each system component easy to be consumed (e.g., Planner as a Service, Orchestrator as a Service, etc.). More information on architecture patterns and differences among context-aware smart environments can be found in (Degeler and Lazovik 2013).

Besides related architectures, we present work related to the algorithms that are being utilized within our architecture as part of the Reasoning Layer (Decision Making component). More specifically, we present the state of the art of scheduling of cloud resources and applications on clouds, as well as automated planning to compose web services and cloud application, and to compare it to present work. The problem of scheduling of cloud resources has been addressed in a number of papers. Cloud scheduler described in (Armstrong et al. 2010) manages user-customized virtual machines in response to a user's job submissions. Its main motivation is to provide computing resources to the research community. Similarly, in (Kim et al. 2010), solution is oriented toward the same application area by providing a scheduling scheme for scientific applications which require large-scale computing resource for long term execution period. Contrary to this, the motivation of our work related to the scheduler is to provide a scheduling mechanism for requests for cloud resources, being used in actual operating environments. Moreover, our scheduler guarantees optimality of schedule in regard to number of used resources for a defined interval of time, that was not tackled neither in (Armstrong et al. 2010) nor in (Kim et al. 2010). In (Lu and Gu 2011), different metrics such as the change of load are used to dynamically schedule cloud resources. By real-time monitoring of performance parameters of virtual machine, scheduling of cloud resources is being done using *ant colony* algorithm to bear some load on load-free node. On the other hand, our scheduler as an input has user-defined metrics, such as resources specification, requested usage duration and policies.

   Scheduling of grid applications on clouds is presented in (Chaves et al. 2010) where not only resource demands are taken into account, but also software requirements of the applications. This approach is similar to ours in sense of taking a content of resources into consideration. Difference is that our approach focuses on service-oriented systems, whereas in (Chaves et al. 2010) they are using grid application of image processing. Besides that, we introduce dependencies among services and the way to manage them. Work done in (Ge and Wei 2010) proposes a scheduler which makes scheduling decision by evaluation the entire group of tasks in job queue. The preliminary simulation results show that scheduler can get shorter "make span" for jobs and achieve better balanced load across all the nodes in the cloud. Instead, our scheduler enables control of maximum number of used resources per a given interval of time. Additionally, there are two papers, (Larsson et al. 2011) and (Sotiriadis et al. 2011), which are focusing on inter-cloud scheduling (scheduling for cloud federations). This represents a different problem, but both papers provide useful insight into specifications, scheduling, and monitoring of services. There are also industry white papers that present how usage of cloud resources can support Agile Software development (CollabNet 2011), (Amit Dumbre 2011). The main idea of these papers is that realization of automated builds, testing and production deployment in clouds can accelerate feedback mechanism that is crucial for Agile software development methodology.

   We review the work related to automated composition of software applications that spans both the AI and cloud computing communities. Many studies use automated planning to compose Web services (*e.g.*, (Kaldeli et al. 2011)), and to automatically generate information flows (Riabov and Liu 2005, Sohrabi et al. 2013), which is a similar problem to Web service composition. Among those studies, HTN planning is employed to represent and compose Web services in multiple approaches (Georgievski and Aiello 2015). The most common one translates the service knowledge from Web Ontology Language for Services (OWL-S) (Martin et al. 2007) to HTNs (Sirin et al. 2004). The main difference between OWL-S and Aeolus lies in that the latter is envisioned for capturing deployment processes of distributed cloud applications, while OWL-S is specifically designed to support the discovery, composition and monitoring of Web services.

   There are also attempts to use automated planning for composing cloud applications. Arshad et al. (2003) describe a deployment problem of software components, and use general-purpose temporal-based planner to find the most optimal plan with respect to plan duration. Lascu et al. (2013) represent a de-

ployment problem using a simplified Aeolus model, and develop a specialised
planner to search for a solution. While the former study does not define the planning problem on any formal ground, we use the simplified Aeolus formal model
as in the latter study to derive our HTN planning problem. Contrary to (Lascu
et al. 2013), where domain-related processes and features are implemented and
embodied in the planning process, we use a general-purpose HTN planner, and
encode the specific knowledge into the domain model.

8.3   Research Projects

We present both implemented and ongoing research projects that relate to energy efficiency in buildings There are several implemented ICT research projects
supporting energy efficiency in buildings, most of which are identified by REEB
project [1], the European Strategic Research Roadmap to ICT enabled Energy-
Efficiency in Building and Construction. In Table 8.2, we refer to the projects
identified by REEB and extend the list with additionally identified projects that
relate to the present work.

Table 8.2: Research projects with ICT support for enegy efficiency

Project title	Period	Website
BuildWise	2007-2010	zuse.ucc.ie/buildwise
AIM	2008-2010	www.ict-aim.eu
BeAware	2008-2010	www.energyawareness.eu
intUBE	2008-2011	zuse.ucc.ie/intube
SM4ALL	2008-2011	sm4all-project.eu
SMOG	2008-2012	n/a (Section 1.3.2)
DEHEMS	2009-2011	www.dehems.eu
eDIANA	2009-2012	www.artemis-ediana.eu
SmartHouse/SmartGrid	2009-2013	smarthouse-smartgrid.eu
GreenerBuildings	2010-2013	greenerbuildings.eu (Section 1.3.3)

The projects listed in Table 8.2 have different focus and objectives. Details on
projects are taken over from the official projects' web pages. *BuildWise* stands
for: Building a Sustainable Future: Wireless Sensor Networks for Energy and
Environment Management in Buildings. "The objective of this project was to
specify, design, and validate a data management technology platform that supports integrated energy and environmental management in buildings utilizing a

---

[1]http://cordis.europa.eu/project/rcn/86724_en.html

combination of wireless sensor network technologies, an integrated data model and data mining methods and technologies. The project supports the development of an integrated software tool for the design and deployment of wireless sensor networks for buildings, including power, signal strength, network protocols and interfaces to existing building management systems." Even though the project deliverables were not available online, for the description we conclude what the differences are. Difference between this project and our project is that we apply automated control whilst they mostly focus on network protocols and interfaces to existing building management systems.

AIM's main objective is to foster a harmonized technology for profiling and managing the energy consumption of appliances at home. "AIM introduces energy monitoring and management mechanisms in the home network and will provide a proper service creation environment to serve virtualisation of energy consumption, with the final aim of offering users a number of standalone and operator services. Behind this goal, the main idea is to forge a generalized method for managing the power consumption of devices that are either powered on or in stand-by state. Especially for the second category of devices, the project will define intelligent mechanisms for stand-by state detection, using all-device-fit control interfaces. The AIM technology is applied on white goods (refrigerators, kitchens, washing machines, driers), communication devices (cordless phones and wireless communication devices for domestic use) and audiovisual equipment (TV Sets and Set-top-boxes)." As it can be seen, the AIM project focuses on the energy consumption of appliances at home, while the focus of our project are non-residential buildings. This difference in application domain, leads to different devices to be controlled, as well as different use cases, scenarios and algorithms to be applied.

"BeAware studies how ubiquitous information can turn energy consumers into active players by developing: (1) an open and capillary infrastructure sensing wirelessly energy consumption at appliance level in the home; (2) ambient and mobile interaction to integrate energy use profiles into users everyday life; and (3) value added service platforms and models where consumers can act on ubiquitous energy information and energy producers and other stakeholders gain new business opportunities. Through this project a number of scientific papers were published. The publications present among others a framework for residential services on energy awareness, designing effective feedback of electricity consumption for mobile user interfaces, as well as increasing residential energy awareness with disaggregated real-time feedback. BeAware has created solutions to motivate and empower citizens to become active energy consumers, by offering them the opportunity to raise awareness of their own power con-

sumption in real time. These solutions include: (1) Energy Life: a mobile phone application, (2) Watt-Lite Twist: an ambient interface that makes use of the home lighting and lamps as a means to communicate with the user, (3) Service Layer: a solution providing data to the interfaces, and (4) Sensor Layer: a solution containing sensor network installed in the users homes and data storage and handling massive amounts of recorded data." Same as the previous project, the BeAware project has households as the application domain and it works specifically on sensor networks, data storage and interfaces for end-users. Unlike the Sustainable Bernoulliborg project, it does not have any form of automation deployed.

As presented in Section 1.3.2, the aim of the *SMOG project* was to introduce energy consumption metering in the municipality of Groningen (NL). An application collecting and presenting energy usage data was developed. By the end of the project, 10 buildings, comprising office buildings, schools, a waste management site and a pump station, were connected to the energy data monitoring and displaying system. The main goal of this project was to collect and present energy consumption data to the end users, to motivate them to perform energy saving actions. Similarity of this project with our project is that both projects deal with data collection, storage and visual representation. However, for the SMOG, monitoring was the core idea, while for the Sustainable Bernoulliborg it is only part of a layer where the system interfaces with the end users, while main part of the system lays in sensing, decision making and control (i.e., automation).

intUBE stands for Intelligent Use of Buildings' Energy Information. As stated by the project leaders: "intUBE promotes increased life-cycle energy efficiency of the buildings without compromising the comfort or performance of the buildings by integrating the latest developments in ICT-field into Intelligent Building and Neighborhood Management Systems and by presenting new ICT-enabled business models for energy-information related service provision. Even though there is little materials that could be found on this project, leaders of the intUBE project promised the following. IntUBE develops tools for measuring and analyzing building energy profiles based on user comfort needs. These offer efficient solutions for better use and management of energy use within buildings over their life-cycles. Intelligent Building Management Systems will be developed to enable real-time monitoring of energy use and optimization. They will, through interactive visualization of energy use, offer solutions for user comfort maximization and energy use optimization. Neighborhood Management Systems is developed to support efficient energy distribution across groups of buildings. These support timely and optimal energy transfers from building to building based on user needs and requirements. New Business Models to make best use of the developed Management Systems are created. The results of IntUBE are expected to enhance not only the comfort levels of building users, but to also reduce overall energy costs through better energy efficiency. These results will be demonstrated in at least three pilot cases: social housing in Spain, office buildings in Finland and a third case defined during the project." From the available materials it is not clear to which degree the promised parts of the project were achieved. However, the main difference of this project comparing to our project is that it focuses on tools for measuring and analyzing building energy profiles based on user comfort and does not explicitly mention any form of automation.

"SM4All has studied and developed an innovative platform, based on a service-oriented approach and composition techniques, for smart embedded services in immersive environments. This has been applied to the challenging scenario of private homes having inhabitants with diverse abilities and needs (eg young, elderly or disabled people). The SM4All project has the ambitious goal of boosting and structuring the cyber intelligence surrounding us in order to simplify our lives. The basic idea is to bring together all devices present in a house and coordinate their activities automatically in order to execute complex tasks that involve many appliances (such as preparing a bath, creating a certain mood in a room, following a video, saving energy, closing the house, etc.). Inhabitants can both interact with and programme the intelligent house, in a simple fashion, through user devices such as the iPad and smartphones. Demos have been built using Brain Computer Interface technology, which allows the user to interact with the SM4all environment using brain waves, without touching any input device. The user simply concentrates on a specific icon shown on a screen in order to initiate an action." (Mecella and Baldoni 2011) Main differences between SM4All and our project are application areas, system goals target users and their interaction with the system. SM4All focuses on private homes while we are focusing on office buildings. Furthermore, they include specific groups of people, such as elderly or disabled people, while we focus on university employees. Additionally, SM4All has the main goal to simplify lives of their users, while we mainly focus on reducing energy consumption while preserving the level of comfort. Finally, the end-users of SM4All interact with homes using different devices, while we strive to understand intentions of users by letting them use building systems the way they are used to.

DEHEMS stands for the Digital Environment Home Energy Management System. This project is looking at how technology can improve domestic energy efficiency. "DEHEMS aims to extend the current state of the art in intelligent meters, moving beyond energy *input* models that monitor the levels of energy being used to an *energy performance model* that also looks at the way in which the energy is used. Bringing together sensor data in areas such as household heat loss and appliance performance as well as energy usage monitoring, it offers real time information on emissions and the energy performance of appliances and services. In turn the potential exists to make changes to appliances/services remotely from the mobile phone or PC. The system can also provide specific energy efficiency recommendations for the household. The potential is to personalize action on climate change, and so help enable new policies such as Personal Carbon Allowances as well as supporting the move towards increased localized generation and distribution of energy. This project resulted in more than 20 scientific publications as well as many project deliverables published. It mostly focused on smart metering and had a number of Living Labs. The Dehems Living Labs are in Birmingham, Bristol and Manchester and in Plovdiv and Ivanovo in Bulgaria. In Dehems the living lab volunteers help us gain insights into peoples behavior regarding energy savings. Feedback from users enabled to develop a customer friendly system that many more people will like to use." Similarly to previous projects, this project focuses on households and does not primarily focus on automation.

eDIANA stands for Embedded Systems for Energy Efficient Buildings. It addresses the need of achieving energy efficiency in buildings through innovative solutions based on embedded systems. "The technology developed in eDIANA improves energy efficiency and optimize buildings energy consumption, providing real-time measurement, integration and control. Moreover, comfort is improved by making the user aware and enabling user-controlled policies for household devices (lighting, domestic electronics, etc.). The eDIANA Platform is a reference model-based architecture, implemented through an open middleware including specifications, design methods, tools, standards, and procedures for platform validation and verification. eDIANA Platform enables the interoperability of heterogeneous devices, and it provides the hook to connect the building as a node in the producer/consumer electrical grid. Thus, this project provides a Reference Architecture for a network of composable, interoperable and layered embedded systems that will be instantiated to several physical architectures. The eDIANA Platform realizations cope with a variable set of location and building specific constraints, related with parameters such as climate, configuration (one to many, one to one etc), energy regulations etc." Difference between eDiana and Sustainable Bernoulliborg projects is that besides the providing the integrated infrastructure, we also sense changes in the environment and we take user behavior into account to improve automated control. Moreover, we have a permanent installation in an operating environment occupied with real end-users, while other mentioned projects are mostly focused on temporary home or living lab installations.

"The *SmartHouse/SmartGrid* project sets out to validate and test how ICT-enabled collaborative technical-commercial aggregations of Smart Houses provide an essential step to achieve the needed radically higher levels of energy efficiency in Europe. This project focuses on both monitoring and control for energy efficiency in the smart homes as well as integrating of smart homes with the smart grid through web services. Furthermore, it tackles multi-agent coordination in the electricity grid together with simulation of a smart grid city with software agents." This project focuses on homes and residential users, and therefore have different use cases and equipment to be monitored and controlled.

"*GreenerBuildings* develops an integrated solution for energy-aware adaptation of public buildings. It investigates self-powered sensors and actuators, occupant activity and behaviour inference, and an embedded software for coordinating thousands of smart objects with the goals of energy saving and user support. GreenerBuildings embraces the following key principles in order to achieve its goals: living-lab experimentation and validation, an agile interdisciplinary consortium, and a user centric approach. In particular, the validation will consider test cases with at least 1.000 networked devices deployed in living-lab buildings." Difference between GreenerBuildings project and Sustainable Bernoulliborg project can be seen as evolution and moving from theory to practice (details described in Section 1.3.3).

Our work is linked to several other ongoing or recently completed research projects within FP7/Horizon 2020 programmes and CIP programmes related to the impact of ICT for Energy Efficiency in Buildings. Also, we build on insight derived from recent FP7/Horizon 2020 programmes focusing on social and behaviour aspects.

More than 100 ongoing or completed projects were found searching the CORDIS web page with the search term "ict for energy efficient buildings". Table 8.3 summarizes the basic information of these (recent or ongoing) projects relevant for present work.

Here we discuss some ongoing projects that are the most relevant to our work, their descriptions, and relations to the our project. Since most of listed projects are still ongoing, we closely follow their work through web pages and personal contacts. Mentioned projects strive to increase energy saving motivation by monitoring consumption and behaviors and providing feedback to raise awareness.

ENTROPY stands for Design of an Innovative Energy-Aware IT Ecosystem for Motivating Behavioral Changes towards the Adoption of Energy Efficient

Lifestyles. ENTROPY project aims at addressing innovative solutions for energy efficiency improvements by understanding the main energy consuming factors and trends, as well as properly modeling and understanding the citizens' behavior and the potential for lifestyle changes. Novel practices that fully integrate information collected from a set of sensor networks and mobile crowd sensing activities are going to be exploited along with processes for monitoring, reporting and analyzing sets of data with regards to energy consumption and the behavioral profile of citizens. The designed IT ecosystem is planned to be validated in three pilot sites. This project focuses on data collection using sensor networks as well as mobile crowd sensing. As stated, the collected data will be used to gain better understanding of user behavior and how it affects consumption. This project does not aim to have automation of control of energy consuming devices.

OrbEEt stands for ORganizational Behaviour improvement for Energy Efficient adminisTrative public offices. This project is aiming at the development of an IT ecosystem for enhanced energy performance monitoring and the display of energy certificates. Users will be engaged with IT tools through intrinsic/extrinsic human motivators. The validation is foreseen in 4 public buildings in 4 countries. The action aims at triggering a 20% average energy demand reduction per building, a 30% CO_2 emission reduction with a 2 years payback period of the tool. This project sets ambitious goals for energy demand reduction and payback period of the system. Our experience showed that behavior change may be rather difficult factor to be controlled.

EnerGAware is acronym of Energy Game for Awareness of energy efficiency in social housing communities. This project is combining the use of gaming, social networking and personalized data driven engagement in energy efficiency. The social media features will provide users with a platform to share data of their achievements, compete with each other, give energy advice, as well as to form virtual energy communities via which the users can learn to balance the energy consumption, comfort and financial cost of their actions.

GreenPlay is the Game to promote energy efficiency actions. This project is aiming at raising awareness among citizens through the implementation of a real time platform. This system will consist of a web-based platform to monitor the energy consumption in real time. Also, advice and challenges to reduce consumption will be available for users of the platform, and a serious game to raise awareness is foreseen. The pilots will be 200 social housing in three different countries.

TRIBE stands for TRaIning Behaviours towards Energy efficiency: Play it! The project aims at the development of a social game to engage users and trigger behavioral change towards energy efficiency, by overcoming non-technical barriers. The social game will be supported by an initial energy audit and diagnosis, virtual pilots in conformity with the 5 real buildings, a funding scheme merging existing instruments with clean web solutions and a user engagement campaign addressing the specific behavior change challenges.

Comparing to all above-mentioned five projects, our project includes (1) developing novel models by employing artificial intelligence algorithms and methods, (2) considering acceptability of energy efficiency solutions, and (3) introducing community-centered design, in addition to user-centered approach, for energy efficiency solutions.

The next six ongoing projects are funded by European Commission and other funding agencies in USA, Germany, and Sweden). However, these six projects do not target publicly owned buildings (including administrative offices, social housing, and buildings in public use or of public interest).

UMBRELLA stands for Business Model Innovation for High Performance Buildings Supported by Whole Life Optimisation. UMBRELLA addresses the energy efficiency through the development of an innovative, web-based decision-support application, which provides common independent evaluation tools built around new and adaptable business models. The interface will use guided navigation to ascertain key information from key stakeholders, building location, building type, owner objectives and preferences, carbon and budget requirements etc. Business models, specific to the project and stakeholders will then be provided through the online dynamic web portal, which will allow users to explore and optimize different business models and the relating implications and recommendations for interventions to their specified building.

DAREED is the acronym for Decision support Advisor for innovative business models and user engagement for smart Energy Efficient Districts. DAREED aims at delivering an ICT service platform (and some specific tools) to foster energy efficiency and low carbon activities at neighborhood, city and district level. Project results will be validated via pilots in three different countries and contexts, thus granting the possibility to generalize results and ensuring replicability throughout Europe and beyond. The key success factor for effective energy efficiency initiatives at community level is to involve all the stakeholders who have active role in decision making and provide them the right information at the right time to take informed decisions; to this extent, user engagement through social networks can foster participation and energy consumption awareness.

IMPRESS stands for Intelligent System Development Platform for Intelligent and Sustainable Society. The aim of the IMPRESS project is to provide a Systems Development Platform which enables rapid and cost effective development of mixed criticality complex systems involving Internet of Things and Services and at the same time facilitates the interplay with users and external systems. The IMPRESS development platform will be usable for any system intended to embrace a smarter society. The demonstration and evaluation of the IMPRESS platform will focus on energy efficiency systems addressing the reduction of energy usage and CO_2 footprint in public buildings, enhancing the intelligence of monitoring and control systems as well as stimulating user energy awareness. This project has similar goals to our implemented GreenMind system. Comparison of these two project may be beneficial for system architects designing systems in the area of smart energy spaces.

In the project *Active House in the Sustainable City*, at the end of 2012, a test family moved into a prototype flat a living lab in the new urban area Stockholm Royal Seaport. The project aims at addressing automatic systems and new mobile tools which will enable the family to keep track of their electricity consumption and help them make sure they use electrical devices when the electricity is at its cheapest and produced in the most environmentally friendly way. The research project involves participants from a broad spectrum of industry actors and academia.

Similarly to our work on *Energy Competition Dashboard* presented in Section 4.3.1,*Power House* project develops "Power House" - a multi-player game - to study how games can be used to build energy efficiency habits. In the field study participants played the game in their homes over the course of one week to one month while their smart meter provided home energy consumption data for analysis. Participants in this study would typically play the game within a real social context. For example, while playing the game via Facebook, players were able to post in-game achievements and energy savings for their Facebook friends to see.

IT4SE stands for IT for Smart renewable Energy generation and use. This is bilateral German-New Zealand project aiming at maximizing generation of cost-effective renewable energy, and its conservation through more efficient use. Information technologies, including novel mobile and web-based services, will motivate and empower citizens to take an active part in this endeavor. This project aims to develop solutions similar to our implemented web and mobile solutions, presented in Section 3.2.4.

8.4 Commercial Products and Services

There are many commercially available products or services supporting energy efficiency in buildings. However, for the illustration, we list a few popular products that relate to energy consuming systems that we include as a part of

Title	Acronym	Period	
Building Energy Efficiency for Massive market Uptake	BEEM-UP	2011	2014
Novel Business model generator for Energy Efficiency in construction and retrofitting	NEWBEE	2012	2015
Holistic Platform Design for Smart Buildings of the Future Internet	HOBNET	2010	2013
Knowledge-based energy management for public buildings through holistic information modelling and 3D visualization	KNOHOLEM	2011	2014
Retrofitting Solutions and Services for the enhancement of Energy Efficiency in Public Edification	RESSEEPE	2013	2017
Intelligent Services for Energy-Efficient Design and Life Cycle Simulation	ISES	2011	2014
Decision support Advisor for innovative business models and user engagement for smart Energy Efficient Districts	DAREED	2013	2016
Energy Efficiency and Risk Management in Public Buildings	ENRIMA	2010	2014
Buildings Energy Advanced Management System	BEAMS	2011	2014
Energy management and decision support systems for Energy Positive neighborhoods	EEPOS	2012	2015
Future Internet Smart Utility Services	FINESCE	2013	2015
Energy Efficiency for EU Historic Districts Sustainability	EFFESUS	2012	2016
Sustainable Zero Carbon ECO-Town Developments Improving Quality of Life across EU - ECO-Life	ECO-LIFE	2010	2015
Energy Forecasting	NRG4CAST	2012	2015
New systems, technologies and operation models based on ICTs for the management of energy positive and proactive neighborhoods	E[PLUS]	2012	2016
Building Energy decision Support systems for Smart cities	BESOS	2013	2016
Control and Optimisation for Energy Positive Neighborhoods	COOPERATE	2012	2015
BARriers for ENERGY changes among consumers and households	BARENERGY	2008	2012
Low Carbon at Work	LOCAW	2011	2014

Table 8.3: National or international research and innovation activities

control system described in this thesis, namely lighting, appliances, heating and air conditioning. In Table 8.4, we present five products that enable monitoring and control of different energy consuming subsystems.[2]

Table 8.4: Energy management products and services available on the market

Product name	Short description	Website
Plum	Scheduiled light control	www.plumlife.com
Plugwise - Circle	Energy meter & switch	www.plugwise.com
NEST	Learning Thermostat	www.nest.com
Plugwise - CoolDing	Airco remote controller	www.plugwise.com
en:Key	Heating control	http://www.enkey.de

The products described in Table 8.4 provide different functionalities. The *Plum Lightpad Dimmer* is a light switch controllable by swipe movement that detects presence and adjusts lights accordingly. A single Plum Lightpad replaces a switch or dimmer in a house and controls the lights that are connected to that switch. If a family room downlight switch is replaced with a Plum lightpad, it is possible to turn on and off or dim those downlights from a smartphone or from the Lightpad itself using a gesture.

The *Plugwise Circle* measures the energy usage of connected devices and lights and allows putting together a schedule to switch devices on and off automatically. The information is passed through the wireless ZigBee-Mesh network to a Plugwise application or to Plugwise software. This product enables near real-time measuring of power consumption and manual and schedule-based control of devices connected to the Circle.

The Google's *Learning Thermostat NEST* is an electronic, programmable, and self-learning Wi-Fi-enabled thermostat that optimizes heating and cooling of homes and businesses to conserve energy. It takes user actions into account to learn from them and generate schedules for automated control.

The Plugwise company also provides the *CoolDing* product which acts are remote controller for air conditioning. CoolDing is a universal infrared remote control for air conditioners (HVAC). CoolDing can record the commands of the remote control of your air conditioner. When recorded, you can play these commands using the CoolDing application on a smartphone or tablet. This makes it possible to control air conditioner from anywhere. It is also possible to automatically switch on/off the AC and adjust the temperature by using a schedule which can be manually programmed.

---

[2]The presented information is acquired from the official websites of the listed companies.

The *en:Key* product represents combination of room sensor and fully automatic valve controller. According to field experiences, it saves up to 20 percent of heating energy. Room sensor and valve controllers are easy to install, without any noise and dirt. The two components do not need any cables for communication or for power supply, with support of to wireless technology, the solar module and the thermal generator. For operation, they produce the necessary energy directly from sunlight and the heat from hot water. en:Key is provided by Kieback&Peter[3] company. This company provides comfort in rooms and maximized energy efficiency. Using networked technology they optimize the interaction of heating, ventilation, air conditioning, lighting, blinds and other systems. Kieback&Peter is globally positioned, with numerous international branch offices, customers and references.

The mentioned products focus on separate subsystems, from heating and air conditioning to lighting and plug load. Moreover, they provide manual, schedule-based and semi-automated control. Our proposed system builds on those solutions, with the aim of being hardware-agnostic. It integrates several types of energy consuming subsystems together and that way increases accuracy of detection algorithms and maximizes energy saving. Moreover, our system is based on AI algorithms that enable it to be completely automated and to act reactively to changes in a building environment, instead of solely learning from user actions.

When it comes to commercially available services, taking into account that the system described in this thesis could be provided as a service, we observe several companies that provide similar services.

Lucid[4] is an American company. Lucid provides energy dashboard for both managers and for public. From the Lucid website we understood that their dashboard can be used to influence user behavior, thus bring savings. However, to the best of our knowledge currently they provide no automated control on energy consuming devices. On the contrary, we provide not only monitoring, but also automated control solutions.

Linc[5] claims to provide hardware and software that enables real-time energy management in buildings, and to provide actionable insights and analytics to help reduce energy consumption and carbon emissions by up to 30%. Linc is a Denmark-based startup company. The differences here are (1) we don't provide hardware solutions - we are hardware agnostic, (2) we provide automated control, and (3) we provide occupant dashboard.

---

[3] http://www.kieback-peter.de
[4] https://www.lucidconnects.com
[5] http://www.linc.world

Wattics[6] is a company from Ireland. Wattics provides an energy measurement analysis tool without automated control solution, while we use advanced monitoring and intelligent automated control to bring additional energy savings.

Another company from The Netherlands is *QwikSense*[7]. QwikSense produces both hardware and software. Our focus, unlike QwikSense, is mainly on software that is compatible with different hardware solutions (device-agnostic). QwikSense also does not offer automated control solution.The differences with our proposed system is that are (1) we are hardware agnostic, (2) we provide automated control, and (3) we provide occupant dashboard.

Besides these companies that are providing disruptive solutions, currently, the market is dominated by large multinationals, such as: *Siemens, Honeywell, Johnson Controls,* and *Schneider Electric.*

Siemens[8] provides building automation systems for all building types and sizes. Siemens uses open communication standards and interfaces to integrate a wide choice of different building control disciplines like heating, ventilation, air conditioning applications, lights and blinds, up to safety features, and equipment. With special features such as centralized, energy management and efficient and innovative energy saving functions, and the effective interaction of all system components and processes they claim to achieve significant cost and energy savings.

Honeywell[9] is a Fortune 100 company, which solutions to improve the quality of life of people around the globe: generating clean, healthy energy and using it more efficiently. SHoneywell offers are wide range of products, such as: Building Automation Systems and Controls, Smart Building Control Systems, Lighting Controls, Building Air Quality Control, etc. For instance, Honeywell Lighting Controls offer smart scheduling, presence detection and photo sensing to ensure lights are not left burning in vacant areas and are off in areas where there is plenty of natural light. It all adds up to energy savings as a standalone solution or as part of an overall integrated energy management plan using heating, ventilation and air conditioning (HVAC) and building automation systems (BAS).

Johnson Controls[10] is a global diversified technology and industrial company serving customers in more than 150 countries. They offer products, services and solutions to optimize energy and operational efficiencies of buildings; lead-acid automotive batteries and advanced batteries for hybrid and electric vehicles; and

---

[6]http://www.wattics.com
[7]http://www.qwiksense.com
[8]http://www.siemens.com
[9]http://honeywell.com
[10]http://www.johnsoncontrols.com

seating components and systems for automobiles. Their Building Automation System features a user interface to better manage HVAC, lighting, security and fire protection systems on a single platform. Enhanced system configurations, programming capabilities, and additional IT features improve productivity, reduce energy costs and enhance security.

Schneider Electric[11] develops technologies and solutions to make energy safe, reliable, efficient, productive and green. The Group invests in R&D in order to sustain innovation and differentiation, with a strong commitment to sustainable development. Their building automation system solution (SmartStruxure) delivers the right information customizable to customers needs on an attractive interface. Schneider Electric claims that day-to-day operations are significantly easier with drag-and-drop trending, calendar-like schedules and one-click reporting. Native open protocols provide the freedom to choose the right equipment for unique application. SmartStruxure solution can deliver deliver an efficient enterprise with up to 30% energy savings while creating a healthy and sustainable environment.

A common thing for all these companies is that they are large multinational companies focusing on global market. That brings certain advantages as the market gets is large. However, it also leads to reduced attention that can be provided to each customer and in many of cases that leads to systems not totally adjusted to reduce energy consumption as much as it's possible. That is where our proposed system comes into play. Moreover, they offer total integrated solution that is expensive and has to be installed with the help of professionals. Comparing to their offer, our presented solution is cheaper and can be installed by any installation company. Our solution is also offered as a service (on demand) and it can be used only for time that is actually needed. Moreover, it is more affordable by any building in need for monitoring and automated control. Our system also leaves the possibility to integrate with the building automation systems that use open protocols, and that way reuse the existing installed infrastructure in combination with improved control algorithms, to provide full automated reactive control.

---

[11]http://www.schneider-electric.com

Chapter 9

Conclusions

We set out to explore how an ICT system can contribute towards more sustainable buildings and to identify what should be the characteristics of such a system, the possible energy and other resource savings resulting from use of different interventions and techniques, and the factors affecting economic and social acceptability.

Evaluations of environmental savings, economic and social acceptability for each proposed intervention showed that the proposed system is acceptable for the stakeholders. As presented in the Figure 9.1, the lowest environmental savings of the interventions were 5%, while the highest were in the order of 80%. These percentages are specific for the environments where the interventions were applied. In any case, each realized intervention showed environmental savings.

The payback period of each intervention falls within 10 years, which was also one of the original goals of the project. More precisely, the payback of each solution varies from 0 to 7,81 years. Moreover, this payback period was confirmed to be acceptable by facility managers who were direct advisors of decision makers (i.e., managing directors).

Most of the interventions were regarded as highly acceptable by the end-users, while one was considered to be of medium acceptability. One solution (Computer control) was not completely evaluated for its user acceptability as there were insufficient survey responses by end-users to allow for statistically relevant conclusions. Even though most of the results indicate high user acceptance, the surveying process made us aware of several important points to be improved in future research.

In order to evaluate user satisfaction with the overall system, we conducted an end-user satisfaction survey, shown in Figure 9.1. In total, 63 building occupants responded to the survey. The results of the survey show that the majority of survey participants (63,49%) were satisfied with the overall system. A smaller percentage of participants (20,63%) gave a neutral mark, while 15,87% were not satisfied.

Finally, we surveyed energy and facility managers as the main stakeholders

	Savings	Payback period	User acceptability
Lighting control	80%	7,81 years	High (77.22%)
Computer control	9,46%	0 years	n/a
Sensor holder	28,34%	3,45 years	Medium (55,56%)
Waste separation	21,73%	4,66 years	High (79,43%)
Water reduction	5%	1,99 years	High (74,6%)

Table 9.1: Summary of the evaluation of the interventions

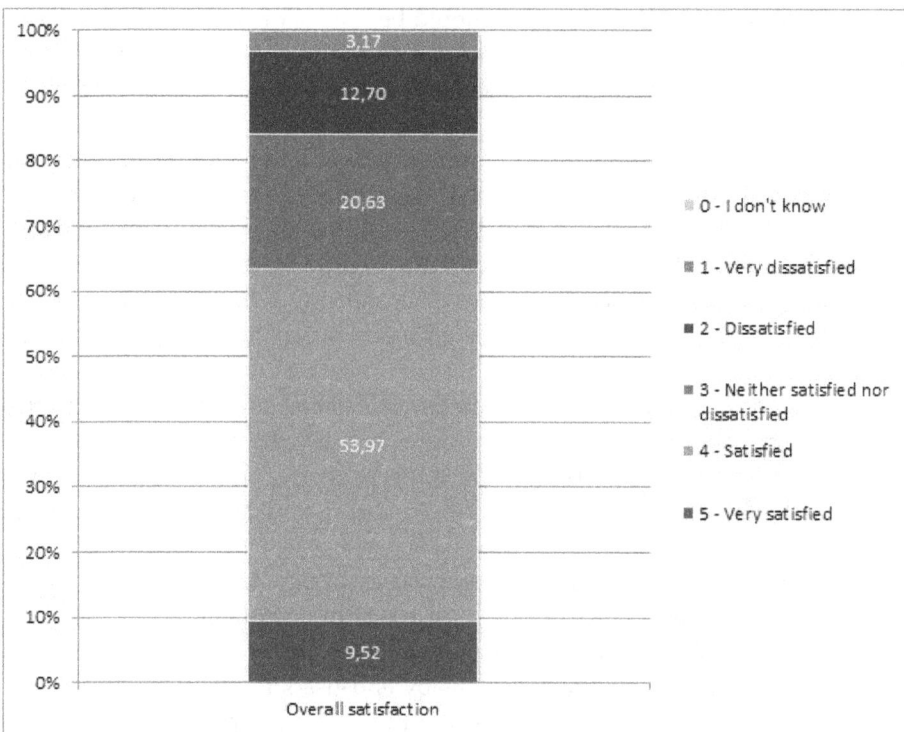

Figure 9.1: End-user satisfaction survey

of the system. On a scale of 1 to 7, they stated that they found this particular system to be easy to use (*perceived ease of use*: 5,17 out of 7) and that they believed the system would enhance their job performance (*perceived usefulness*: 5,67 out of 7). Moreover, all subjects interviewed stated that they would be interested in using the system in buildings which they manage (*intention of use*).

9.1   System Requirements Satisfaction

Considering the original system requirements presented in Section 2.2.2, we qualitatively evaluate to which extent these are satisfied by the system. Moreover, we identify the less urgent requirements that are partially satisfied and left for future work.

In total, there were 12 functional requirements. Functional requirements *FR1*, *FR2* and *FR3* are related to environmental conditions and consumption data collection and storage. These requirements were satisfied, as well as implemented and deployed in the actual environment, as described in the Section 4.3.3.

The functional requirements *FR4*, *FR7* and *FR8* relate to user interfaces which visualize the consumption and environmental data. These were also implemented, tested, and deployed in the actual environment. More information about these actions can be found in the Sections 4.3.2 and 4.3.1. The difference between the web application for managers and the one for building users is that the managers have additional functionalities, such as correlating consumption data with environmental data (e.g., weather). Moreover, functional requirement *FR5* (notification system), was also implemented as part of the work described in the Section 4.3.1, while the report generation feature, functional requirement *FR6*, was implemented as a part of the work described in the Section 3.3.4.

Table 9.2: The results of the requirements evaluation

Functional requirements	Evaluation result
FR1 - Environment condition data collection	Implemented
FR2 - Consumption data collection and storage (derived)	Implemented
FR3 - Historic data collection	Implemented
FR4 - Web application for Managers	Implemented
FR5 - Notification system for Managers	Implemented
FR6 - Report generation for Managers	Implemented
FR7 - Monitoring personal consumption for building users	Implemented
FR8 - Monitoring overall consumption for building users	Implemented
FR9 - Control interface to Managers	Dev. in progress
FR10 - Automated control of HVAC system	Dev. in progress
FR11 - Automated control of Lighting system	Implemented
FR12 - Automated control of Appliances	Implemented

Functional requirements *FR11* and *FR12* relate to the control of lighting and appliances. Both features were developed and deployed as a part of work presented in the Section 3.2.3.

The only remaining functional requirements whose development is still in progress are Control interface for Managers (*FR9*) and Automated HVAC control (*FR10*). These requirements were internally marked as less urgent. The reason for this was that the control interface feature came from the facility managers as an additional feature. Moreover, the feature of automated control of the HVAC system heavily depended on the results of automated control of Lighting, which first had to be evaluated before applying the same principles to an HVAC system. For the reasons mentioned, these requirements will be considered as a subject for future work.

There were also 12 non-functional requirements, of which 9 were implemented, and 1 partially implemented; for 2 development is still in progress. The non-functional requirement *NFR1* (GUI simplicity), was covered by involving professional graphical designers that created the graphical user interface design (see Acknowledgments).

The non-functional requirement *NFR2* (Installability) was covered by implementing components of the architecture in the form of micro-services. Using the same technology, fault tolerance (*NFR5*) was partially implemented. Still, there remain some scenarios to be covered; these will be included as subjects for future work.

The non-functional requirement High performance (*NFR6*) was also covered, as the implemented system collects and presents data as well as performing control in real time.

As a part of the PC Sleep Mode control solution (Section 4.3.4), the requirement regarding operational timeouts (*NFR7*) was implemented. Still, as this functionality has not yet been propagated to the whole system we mark it as partially implemented and include it as a part of future work.

The privacy and security requirements (*NFR7, NFR12*) were tackled by not storing personal data about the users in the database, as well by encoding all data used in communication. Data coming from individual sensors cannot be coupled with a particular person. This makes the sensor data truly anonymized.

The non-functional requirement of scalability within a building (*NFR9*) was confirmed by deploying the solution to more sections in the building (see Section 4.1). Portability (NFR11) was confirmed by moving and reusing the data collection solution from one office to the whole floor in the Bernoulliborg. The requirement of scalability among more buildings (*NFR10*) was confirmed by deploying the solution to more buildings of the Municipality of Groningen.

The requirements Configurability (*NFR3*) and Maintainability (*NFR4*) are still in progress, as their implementation on several components has to be spread across the whole implementation. These two requirements are therefore consid-

Table 9.3: The results of the non-functional requirements evaluation

Non-functional requirements	Evaluation result
NFR1 - Simplicity	Implemented
NFR2 - Installability (derived from B1)	Implemented
NFR3 - Configurability (derived from B1)	Development in progress
NFR4 - Maintainability	Development in progress
NFR5 - Fault tolerance	Partially implemented
NFR6 - High performance	Implemented
NFR7 - Performance (timeouts)	Implemented
NFR8 - Privacy	Implemented
NFR9 - Scalability - a building	Implemented
NFR10 - Scalability - more buildings	Implemented
NFR11 - Portability	Implemented
NFR12 - Security	Implemented

ered to be part of future work.

Table 9.4: The results of the business requirements evaluation

Business requirements	Evaluation result
BR1 - Cost effectiveness	*Satisfied*
BR2 - Controllability	*Satisfied*

Finally, the business requirements Cost effectiveness (*BR1*) and Controllability (*BR2*) were satisfied. Cost effectiveness was confirmed by keeping the payback period of the system within bounds considered acceptable by the clients (i.e., within 10 years). Moreover, cost effectiveness will increase once the costs of hardware and other services involved in the initial investment have been reduced. The Controllability requirement was satisfied by preserving all crucial components of the system within the ownership of our development team. Controllability can also be improved by including the expertise of front-end development within the internal development team, which should in turn also positively affect the cost effectiveness.

9.2 Answers to the Research Questions

RQ1: Assuming that a smart energy system is realizable in an actual operating environment, how can real-time consumption data be feasibly obtained and what energy consuming devices can be controlled using an ICT system? What should be the charac-

teristics of an ICT system that supports sustainability measures in buildings?

Real-time consumption data can be obtained in various ways. Depending on where measurements take place, on the main meter or measuring consumption of a device or group of devices, measuring devices such as smart plugs or pulse counters can be used. In our implementation, to measure consumption of an office, an individual user or a single device, we used smart plugs, which communicate using a wireless sensor network and transfer consumption data every one second. For measuring consumption of the whole building we developed, for both electric and optical pulses, a pulse counter that was connected directly to a main electricity meter. Using this solution we were able to count pulses and store or send measurements every second. We also implemented support for collecting signals from the P1 port of a smart meter. Although other methods of metering, such as use of a clamp meter, are possible, for the purpose of this work they were not necessary.

Several types of energy consuming devices can be controlled by using an ICT system. We mostly controlled office equipment such as PCs, laptops, printers, light fixtures, lamps, radios, as well as kitchen equipment like coffee machines, fridges, microwaves, and boilers. Other devices can be also controlled using an ICT system, especially simple devices that are controlled by using a standard electricity switch. For more sophisticated equipment that is not recommended to be simply switched off from the power line, additional hardware or software interfaces need to be used or developed. An example of this is a software interface that we developed to control PCs; instead of wirelessly controlling a smart plug switch, we send a software command to a PC to switch it off or put it into sleep mode.

The characteristics of an ICT system to support sustainability measures in buildings are described in detail in Section 2.2. There we describe functional and non-functional requirements of the system as collected from stakeholders, as well as business requirements collected from case studies. In total, we identified twelve functional, twelve non-functional and two business requirements which are mostly satisfied by the system built.

RQ2: Can an ICT system increase the efficiency of resource use in a building, and if so, which techniques contribute to increase of efficiency while maintaining user comfort?

As a result of our work, we conclude that an ICT system can indeed increase the efficiency of resource use in a building. According to the experiments we performed in the Bernoulliborg, we increased the resource efficiency of five different subsystems by 5-80%. The potential efficiency greatly varies, depending on the spatial and environmental conditions of a building where the system is installed.

Efficiency may be increased by proper understanding of the usage patterns of the space and a proper configuration of the installed system.

The evaluation results show that the most effective technique was the AI planning used for Lighting control intervention. Using this technique, we achieved electricity savings in the order of 80%. Additionally, linear regression proved to be a good technique to discover user preferences for using their workspace and their energy consuming devices. Using linear regression we can obtain constant optimization, and at the same time maintain user satisfaction by including users' negative feedback in the automated control decision process.

RQ3: Assuming that a smart energy system generates economic savings in a building, which factors influence the economic acceptability of such a system? Moreover, which values of those influencing factors are acceptable and realistic?

The studies conducted show that the factor of greatest influence for economic acceptability of a smart energy system was its payback period. Economic acceptability increases as the payback period becomes shorter. Several factors may influence the payback period. As the *Payback period* equals the *Cost of investment* divided by the *Annual saving*, we can reduce the payback period in two ways: (1) by reducing the investment, or (2) by increasing the annual savings. Costs of investment can be reduced by reducing or eliminating some of the variables influencing these costs, such as hardware production, transportation, import taxes, software development, installation costs, and so on. Annual savings can be increased when the efficiency of used software algorithms increases, when the number of working weeks or consumption increases, or when the price of utilities (e.g., electricity, water, etc.) rises.

The actual implementation of the reference project showed that savings could be achieved for each system tackled. Our analysis showed that the acceptable payback period for the system, as defined by stakeholders, falls within 10 years, and in some cases within 15 years. For the systems analysed these payback periods were also shown to be realistic.

RQ4: Is such a smart energy system acceptable for end-users of a building? If so, how important is social acceptability for such a system and which factors influence it?

Social acceptability plays a very important role. A great system that is not used cannot achieve any savings. The system that we developed showed to be acceptable by the building end-users. Measured social acceptability rated from 55,56% to as high as 79,43%.

As noted throughout this work, factors influencing social acceptability of the system are closely connected to the way in which the project goals, motivations and final achievements are communicated to the stakeholders and end-users. Another important factor in social acceptability the amount of extra effort which users must put into the system. Based on the results of the end-users' survey, we conclude that social acceptability could be increased by improving communication regarding a project or individual intervention.

9.3 Sustainable Buildings

T he energy consumption of non-residential buildings around the world contributes significantly to the world's total energy consumption. The work presented in this thesis is driven by the motivation to reduce resource consumption within non-residential (office) buildings using an ICT system, in ways that are efficient, as well as economically and socially acceptable.

In this thesis we used design science research methodology to create a smart energy system intended to make an organization's operations more sustainable. Our work focused on design, implementation and deployment of the system in a real operating environment, as well as optimization of the infrastructure used by the system itself. Moreover, we identified related work and evaluated the system from environmental, economic and social perspectives.

In Chapter 2 we started by collecting the requirements for the system from all involved stakeholders. Through this process, we documented the requirements essential for proper design of the system. We introduced an architecture for a smart energy system, described each component in detail, and explained how the system communicates and operates.

In Chapters 3 and 4 we described how the system is implemented and deployed in the Bernoulliborg in Groningen (The Netherlands). We presented and motivated our technological choices, and described the setting and management of each intervention.

In Chapter 5 we investigated how the use of computing and storage infrastructure can be optimized using AI techniques, scheduling and planning. Moreover, we evaluated the performance of each technique, taking into account a typical size of the resource allocation problems for optimization of building operations.

In Chapter 6 we conducted and presented experiments designed to evaluate efficiency of the system. Moreover, we used real project costs and savings to evaluate the payback periods and economic acceptability of the system.

In Chapter 7 we conducted user surveys to evaluate the social acceptability of the system.

Next, in Chapter 8 we presented our findings regarding work done in similar ICT research projects and energy saving studies, and we described similar software architectures and commercially available products.

Finally, we evaluated the overall system and discussed the factors affecting acceptability. We reviewed satisfaction of the system requirements and provided answers to the research questions.

9.4   Reflection

We have based our conclusions on the implementation of a relatively modern building, the Bernoulliborg. We stipulate, however, that our results are more general and are portable to other office buildings. The solutions for saving energy on PCs are independent of the envelope. The way in which lights are controlled in office buildings or in restaurants, using fixed schedules and PIR, has been the most common one for past decades. So our interventions promise to be easily portable to other buildings and to bring comparable savings. Even when considering a state of the art building like The Edge we think that our integrated solution can bring additional benefits by reusing information between different sensor systems, as well as by using user feedback to correct the control mechanisms. Our current plan to deploy this solution in more buildings of the University of Groningen and several buildings of the Municipality of Groningen will either confirm or invalidate the portability of the solution and transfer of savings. It is also interesting to consider for future research the transfer of these solutions to radically different types of buildings.

The transition to smart and sustainable non-residential buildings is a major change that has to be carried out if we are to reduce energy consumption and overall environmental damage. This transition can go smoothly only if all involved parties understand and embrace the individual and collective benefits of the transition.

It is vital to know what types of interventions can be realized on the way towards more sustainable buildings. Moreover, it is important to understand which interventions have a significant impact, how to evaluate interventions, and which techniques best support the execution of these interventions.

This thesis contributes to the body of knowledge in the area of sustainable buildings by providing a proposal for the design and deployment of a supporting ICT system in an actual operating building. It also contributes to the understanding of crucial factors affecting the economic and social acceptability of such systems.

The system built during this dissertation showed to be scalable and portable to other buildings. That resulted in emerging possibilities and demand for commercial exploitations of the system. The impact of this work can be significant if the presented solutions are extended to support more energy consuming subsystems, and if the system scales out to more non-residential buildings. Moreover, the impact can be increased if the system is also applied to residential buildings, especially for buildings with no building management system. Furthermore, improvements in cost-efficiency over time will make this system more appealing to stakeholders in the residential sector.

9.4.1   Reflection on used methodologies

We reflect on two methodologies used within this thesis: Scrum and Technology Acceptance Model (TAM). In general, the scrum methodology that we used during the software implementation process brought a number of benefits. Most important benefits were reflected in continuous, structured improvement of the products. Each team took responsibility for the delivery and quality of their product(s) under development. The focus moved from traditional project management to realization of the scrum development process. As software was delivered in short iterations, small functional pieces of software were demonstrated frequently, feedback collected from stakeholders, and necessary corrections carried out regularly to the satisfaction of product owners and clients. One drawback of this methodology was that more time was used for team meetings; that placed an additional workload on the developers. However, we mitigated this issue by adjusting the frequency of meetings to the pace of our team and the planned number of developers' working hours; this meant a maximum of three meetings per week.

To evaluate the social acceptability of the GreenMind System, we used the constructs of the Technology Acceptance Model - TAM (Davis 1986). As previously stated, according to TAM a user will have a positive attitude toward using a system if he or she perceives that system to be easy to use as well as useful, and a positive attitude leads to actual system use. In this thesis, we took the TAM method as referent and did not question its validity. However, our observation was that actual users of the system (in our case facility managers or building users) may not necessarily be the ones who decide whether to use or implement the system. The people making those decisions may come from a higher level of management, or even be building owners or building maintainers. Therefore, for future study, besides for gathering requirements, after a project is implemented it would be useful to interview the actual decision makers to understand their reasons for accepting and adopting the system. From our research we understand that decision makers may be more influenced by factors like economic acceptability, and related to this, they may need to know whether a newly installed system would be socially acceptable and therefore used properly to bring the expected savings.

9.5 Future Directions

The issues explored in this thesis are open to further investigation and point to several directions for additional research. Regarding the system itself, there is space for improvement to make the proposed solution more resilient and to increase its compatibility with more hardware components and software interfaces, as well as standard building management systems. Moreover, some characteristics of the system, such as configurability and maintainability, should be improved in order to support its business scalability.

With regard to the Computer Sleep Mode control solution, it would be interesting to compare the results from computers of staff members with the results from publicly available computers for students. The reason is that computers publicly available for students are often left running while not in active use; the sleep mode could in such cases result in even higher savings.

Moreover, the automated control part of the system could be tested with different AI techniques to evaluate which techniques are most effective. Machine learning algorithms supporting dynamic timeouts and personalization of space management should also be applied and tested. For manual and scheduled control, user-interfaces for managers should be implemented.

Additionally, we tried to minimize influences between different saving interventions by implementing them during different time periods. For example, automated lighting control in the restaurant area was carried out during a different period than office PC control, and mainly involved different users. Moreover, all actions that involved informing the building users, such as promotional campaigns using posters, stickers and energy consumption dashboards, were done at the end of the project to minimize the influence of these actions on experiments and measurements performed as a part of other interventions. However, behavioral changes on the part of workers may have had some minor influence on the savings introduced by the technology. In future research, the interventions could be better isolated to minimize influences between interventions. Moreover, social acceptability considerations could be deepened by examining why users are satisfied, or not, and which factors affect differences in satisfaction.

The fact that our team was granted the second Green Mind Award (2014)[1] in a row, should ensure the continuity of this work. The award will enable the

---

[1]http://www.rug.nl/about-us/who-are-we/sustainability/green-mind-award/winners-2014

extension of the system to the one remaining uncovered building subsystems, in particular HVAC. Therefore, as part of this follow up project, a compatible room-level automated heating control solution should be implemented and deployed in one of the buildings of the University of Groningen.

Appendices

Appendix 1 - Sustainable Bernoulli Building - base-line survey (before the start of the project)

The data the researcher collects about you will only be used for this research. The researcher saves your data without a personal information to keep it anonymous. In reports on the study joint results will presented. Participation is completely voluntary.

Q1. How much time in average you spend daily INSIDE your office?

Q2. How much time in average you spend OUTSIDE your office during a working day?

Q3. How much time in average you spend INSIDE your office but not working behind your computer during a working day?

Q4. Which devices do you use during your working hours, and for how long (in hours)?
(Offered items: Desktop computer, Laptop, Additional lamp(s), Radio, Water cooker, Microwave, Coffee machine, Heater, Other device)

Q5. How often you think about energy saving?
□ *Several times everyday* □ *One time a day* □ *Several times a week* □ *One time a week* □ *Several times a month* □ *One time a month* □ *Several times a year* □ *I don't think about it at all* □ *Other*

Q6. Which energy saving actions do you take?
□ *I put my computer to standby mode when going to a lunch break* □ *I put my computer*

to standby mode when leaving my office □ *I turn off my computer when leaving my office*
□ *I set my computer standby timeout to a proper value* □ *I set my computer standby
timeout to a very low value* □ *I do not use any additional devices (besides the computer
and the phone) in my office* □ *Other*

Q7. I often think about the fact that I work in my department
Completely disagree □ □ □ □ □ □ □ *Completely agree*

Q8. I feel committed to other colleagues of my department
Completely disagree □ □ □ □ □ □ □ *Completely agree*

Q9. The fact that I work in my department is an important part of my
identity
Completely disagree □ □ □ □ □ □ □ *Completely agree*

Q10. I feel a bond with other colleagues from my department
Completely disagree □ □ □ □ □ □ □ *Completely agree*

Q11. I have a lot in common with the average colleague of my department
Completely disagree □ □ □ □ □ □ □ *Completely agree*

Q12. I am similar to the average colleague of my department
Completely disagree □ □ □ □ □ □ □ *Completely agree*

Q13. I would feel guilty for consuming a lot of energy at work
Completely disagree □ □ □ □ □ □ □ *Completely agree*

Q14. I would feel proud for conserving energy at work
Completely disagree □ □ □ □ □ □ □ *Completely agree*

Q15. I would feel morally obliged to conserve energy at work
Completely disagree □ □ □ □ □ □ □ *Completely agree*

Q16. People who are close to me would find it important to conserve energy
at work
Completely disagree □ □ □ □ □ □ □ *Completely agree*

Q17. Colleagues from my department would find it important to conserve
energy at work

Completely disagree □ □ □ □ □ □ □ *Completely agree*

Q18. Are there any other aspects that you would like to mention/change in your office?

□ *Intensity of light should be higher* □ *Intensity of light should be lower* □ *A number of lamps should be higher* □ *A number of lamps should be lower* □ *I prefer to have a desk lamp so that I can switch it on/off manually* □ *Sensors that trigger the lights should be set on shorter time-out (lights should stay ON short period of time after there is no movement)* □ *Sensors that trigger the lights should be set on longer time-out (lights should stay ON long period of time after there is no movement)* □ *Sensors that trigger the lights should be placed in a better location inside my room* □ *The room is being overheated* □ *The room is being underheated* □ *Room ventilation is too intensive* □ *Room ventilation is poor* □ *Other*

Q19. If needed, explain desired changes in more detail
....

Q20. Overall satisfaction with the working environment?
Not satisfied at all □ □ □ □ □ □ □ *Extremely satisfied*

Appendix 2 - A Survey for Facility and Energy Managers

Thank you very much for agreeing to be a part of this study! The aim of this interview and survey is to collect information on how facility managers perceive a possible new smart energy system that the university may implement and how they evaluate it. *[The system is introduced]*

Q1. How does this compare with your current building automation systems (BAS), if any?

Q2. What parts of the system could you see your organization adopting? Why?

Q3. What barriers do you foresee can hinder or delay the adoption of such features in a real setting?

Realization of a Smart Energy System
Please indicate to what extent you agree with the statements below if this system would be implemented.

Q4. The system should have the possibility to adjust HEATING, VENTILATION, and AIR CONDITIONING (HVAC) according to presence and/or activity of people inside of a building.
Completely disagree □ □ □ □ □ □ *Completely agree*

Q5. The system should have the possibility to adjust LIGHTING according to presence and/or activity of people inside of a building.
Completely disagree □ □ □ □ □ □ *Completely agree*

Q6. The system should have the possibility to adjust APPLIANCES (PCs, printers, projectors, boilers) according to presence and/or activity of people inside of a building.
Completely disagree □ □ □ □ □ □ *Completely agree*

Q7. The system should have the ability to adjust environment conditions on a level of:
□ *Area (a part of the room)* □ *Section (a set of areas in a larger room, e.g. restaurant)* □ *Room* □ *Wing* □ *Floor* □ *Building* □ *Other*

Q8. The system should be able to scale from the level of one room to the level the whole building.
Completely disagree □ □ □ □ □ □ □ *Completely agree*

Q9. Do you consider it acceptable to have additional hardware add-ons (e.g., sensors, wireless switches, actuators) installed in order to get additional functionalities and/or services within a building.
Completely disagree □ □ □ □ □ □ □ *Completely agree*

Q10. Historic data about the activities and conditions in a building should be stored and reused for building operation optimization purposes (e.g., space usage planning, cleaning, etc.).
Completely disagree □ □ □ □ □ □ □ *Completely agree*

Q11. It is important that the collected data should be owned by the organization providing or generating data.
Completely disagree □ □ □ □ □ □ □ *Completely agree*

Q12. The collected data should be kept by the organization providing the service of smart energy system and has knowledge of secure data collection and storage.
Completely disagree □ □ □ □ □ □ □ *Completely agree*

Q13. In your opinion, this type of a Smart Energy System is most appropriate for:
□ *existing buildings without BAS* □ *existing buildings with BAS* □ *newly built buildings with BAS* □ *newly built buildings without BAS* □ *both existing and newly built buildings with BAS* □ *both existing and newly built buildings without BAS both existing and newly* □ *built buildings with and without BAS* □ *Other*

Q14. A system may NEGATIVELY affect occupant's comfort and productivity.
Completely disagree □ □ □ □ □ □ □ *Completely agree*

Q15. A system may POSITIVELY affect occupant's comfort and productivity.
Completely disagree □ □ □ □ □ □ □ *Completely agree*

Q16. It is acceptable to use the employee public calendar to adjust energy

consuming systems (heating, lighting, appliances, etc.)
Completely disagree □ □ □ □ □ □ □ *Completely agree*

Q17. From the operational point of view, Facility Managers must have the chance to adjust rules to make buildings use optimal.
Completely disagree □ □ □ □ □ □ □ *Completely agree*

Q18. The system should be smart, but the users should have the final decision (e.g., to turn of the light or heating).
Completely disagree □ □ □ □ □ □ □ *Completely agree*

Q19. For building occupants, it would be useful to have user applications (e.g., mobile apps) to MONITOR their PERSONAL energy consumption?
Completely disagree □ □ □ □ □ □ □ *Completely agree*

Q20. For building occupants, it would be useful to have user applications to MONITOR their OVERALL energy consumption of a building?
Completely disagree □ □ □ □ □ □ □ *Completely agree*

Q21. For building occupants, it would be useful to have user applications (e.g., mobile apps) to CONTROL of personal space/office.
Completely disagree □ □ □ □ □ □ □ *Completely agree*

Q22. For Facility managers, it would be useful to have a WEB application to MONITOR the consumption within a building.
Completely disagree □ □ □ □ □ □ □ *Completely agree*

Q23. For Facility managers, it would be useful to have a MOBILE application to MONITOR the consumption within a building.
Completely disagree □ □ □ □ □ □ □ *Completely agree*

Q24. For Facility managers, it would be useful to have a WEB application to CONTROL devices within a building.
Completely disagree □ □ □ □ □ □ □ *Completely agree*

Q25. For Facility managers, it would be useful to have a MOBILE application to CONTROL devices within a building.
Completely disagree □ □ □ □ □ □ □ *Completely agree*

Q26. For Facility managers, it would be useful to receive notifications (e.g., by SMS or email) when the consumption exceeds certain planned or expected limits.
Completely disagree □ □ □ □ □ □ □ *Completely agree*

Q27. For Facility managers, it would be useful to have an option of automated energy REPORT generation?
Completely disagree □ □ □ □ □ □ □ *Completely agree*

Q28. The system should be easy to maintain (e.g., faulty behavior of devices should be detected and maintenance department notified about failures, e.g., broken sensors).
Completely disagree □ □ □ □ □ □ □ *Completely agree*

Q29. The system should have simple interface for users.
Completely disagree □ □ □ □ □ □ □ *Completely agree*

Q30. Privacy of users should not be compromised within this system.
Completely disagree □ □ □ □ □ □ □ *Completely agree*

Q31. The data collected by sensors (e.g., presence sensor) should not be accessible outside the system (e.g., by third-party system)?
Completely disagree □ □ □ □ □ □ □ *Completely agree*

Q32. Users should see clear benefits in order to accept the system (e.g., ability to adjust their system according to their preferences or needs)
Completely disagree □ □ □ □ □ □ □ *Completely agree*

Q33. In a building, the following factors should be measured:
□ *Light levels (for lighting control)* □ *Temperature (for heating and cooling control)* □ *Movement (for presence/absence detection)* □ CO_2 *levels (for the air conditioning control)* □ *Sound (simple microphone for activity detection)* □ *User connection to the WiFi network (for person count)* □ *Other*

Q34. In case of unpredictable failures (e.g., power blackouts) the system should return to its normal working mode.
Completely disagree □ □ □ □ □ □ □ *Completely agree*

Q35. The system should be able to react in the real time.

Completely disagree □ □ □ □ □ □ □ *Completely agree*

Q36. The system should not react too fast, but using some delays/timeouts (e.g., if somebody leaves a room only for a short time the system puts devices in a room to energy saving mode)
Completely disagree □ □ □ □ □ □ □ *Completely agree*

Q37. The system should be reliable. The system should consistently perform according to its functional specifications.
Completely disagree □ □ □ □ □ □ □ *Completely agree*

Q38. The system should not affect the comfort nor productivity of the users, building occupants. In other words, if the the system works properly it should not be noticed by the end-users.
Completely disagree □ □ □ □ □ □ □ *Completely agree*

Q39. In a building, the following factors should be measured:
□ *1-3 years* □ *3-5 years* □ *5-7 years* □ *7-10 years* □ *10-15 years* □ *15-20 years* □ *Payback period is not relevant.*

Q40. In your opinion, maintenance of the system should not be done more frequently than:
□ *once in a week* □ *once in a month* □ *once in 3 months* □ *once in 6 months* □ *once in a year* □ *once in three years* □ *Frequency of maintenance is not relevant.*

Q41. In your opinion, annual maintenance of the system should not be more expensive than:
□ *0.5 fte of a maintenance worker* □ *1.0 fte of a maintenance worker* □ *2.0 fte of a maintenance worker* □ *3.0 fte of a maintenance worker* □ *Other*

Q42. I believe that this particular system would enhance my job performance?
Completely disagree □ □ □ □ □ □ □ *Completely agree*

Q43. I believe that this particular system would be easy to use?
Completely disagree □ □ □ □ □ □ □ *Completely agree*

Q44. Would you be interested in using this smart energy system in your building?

▫ *Yes* ▫ *No* ▫ *Other*

Q45. Would you like to be informed about developments on this project?
▫ *Yes* ▫ *No* ▫ *Other*

Appendix 3 - Questionnaire on new waste separation system

Dear employee of the University of Groningen,

Thank you very much for filling in this questionnaire. The University of Groningen aims to reduce the amount of waste that is produced. One solution would be to install a new waste separation policy in university buildings to increase recycling. Trash bins in offices of employees will be removed and instead, new trash bins will be placed at central locations on each floor. These new trash bins allow for separating paper, can, plastic and other types of garbage. This way, more waste can be recycled, resulting in less waste. The aim of this questionnaire is to test how people working in the Bernoulli building perceive the new waste separation policy and how they evaluate it.

Researchers from the faculty of Behavioral and Social Sciences have drawn up a questionnaire to evaluate the acceptability and effectiveness of this new waste separation system. The data collected from this questionnaire will be treated confidentially, so it wont be possible to deduce individual responses. A summary of the results will be made available. We would like to ask you to fill in the questionnaire.

Completing the questionnaire will take about 10 minutes. By continuing with this questionnaire you indicate the following: you have read the information above, you are aware of the conditions and the procedure of the study as indicated above, you consent to participate in this study.

Q1. What is your gender?
□ *Female* □ *Male*

Q2. Below, you will find a description of a waste separation policy recently implemented by the university to increase recycling among employees of the Bernoulli Building. Please read the policy. *"Recently, all trash bins in your offices were removed. Instead, new trash bins are placed at a central location on each floor, which allows for separating paper, can, plastic and other types of garbage. Hence, you now have to go out of your office every time you need to use the trash bin, and throw your trash to a specific container on your floor. By this, the university aims at increasing the amount of recycled garbage in Bernoulli Building."* Implementation of this policy might have induced certain responses by you. Please indicate the extent to which you responded in the following ways. *(answers on scale 1 to 7)*

□ *1- I protested against the policy* □ *2- I refused to behave in line with the policy* □ *3-*

I accepted the policy □ 4- I felt that the policy was unfair to me □ 5- I agreed with the policy □ 6- The implementation of the policy reduced my productivity at work

Q3. The following statements are about the effort it takes you to unplug devices at home. Please indicate to what extent you agree with these statements: *(answers on scale 1 to 7)*
□ It requires little effort for me to always unplug devices at home □ I unplug devices at home automatically □ I easily forget to unplug devices at home □ It is feasible for me to unplug devices at home □ I am able to unplug devices at home

Q4. Please indicate the extent to which you agree with the following statements. You can answer on a scale from 1 (totally disagree) to 7 (totally agree).
□ 1- The university is considering environmental benefits by implementing such a policy □ 2- The university is considering financial benefits by implementing such a policy □ 3- I trust the decision of the university with implementing this policy

Q5. The following statements are about your recycling behaviour at work. Please indicate to what extent you agree with the statements below on a scale from 1 (never) to 7 (always).
□ I separate my waste at work □ I separate paper from the rest of my waste at work □ I separate cans from the rest of my waste at work □ I separate plastic from the rest of my waste at work

Q6. Please indicate the extent to which you engage in the following behaviours. You can answer on a scale from 1 (Never) to 7 (Always).
□ 1- I recycle paper at home □ 2- I try to reduce the amount of waste at work □ 3- I recycle plastic bottles at home □ 4- I recycle cans at home □ 5- I use my own mug at work instead using disposable cups □ 6- I recycle glass at home □ Several times a year □ I don't think about it at all □ 7- I try to consume products with less waste to reduce the amount of waste at work

Q7. Please indicate the extent to which you agree with the following statements. You can answer on a scale from 1 (totally disagree) to 7 (totally agree).
□ I am the type of person who acts environmentally friendly □ Acting environmentally friendly is an important part of who I am □ I see myself as the type of person who acts environmentally friendly

Q8. Finally we would like to ask you some general questions on what you find important. Below you will find 16 values. Behind each value there is a short explanation concerning the meaning of the value. Could you please rate how important each value is for you?

Your scores can vary from -1 up to 7. The higher the number (-1, 0, 1, 2, 3, 4, 5, 6, 7), the more important the value is as a guiding principle in YOUR life. Try to distinguish as much as possible between your ratings of the values by using different numbers.

□ *equal opportunity for all* □ *harmony with other species* □ *control over others, dominance* □ *joy, gratification of desires* □ *fitting into nature* □ *free of war and conflict* □ *material possessions, money* □ *the right to lead or command* □ *correcting injustice, care for the weak* □ *enjoying food, sex, leisure, etc.* □ *preserving nature* □ *having an impact on people and events* □ *working for the welfare of others* □ *protecting natural resources* □ *doing pleasant things* □ *hard working, aspiring*

Appendix 4 - Smart Lighting in the Bernoulli Building Restaurant

Usability evaluation of the lighting system in the Bernoulliborg restaurant
In the scope of the **Bernoulliborg - The building of sustainability** *project, we have developed and implemented a lighting system that automatically controls the lamps in the restaurant of Bernoulliborg. The restaurant has been equipped with sensors that provide information on, for example, movement of people within a particular space. We have deployed and used the system in the restaurant in the past two weeks. If you have been in the restaurant in these weeks, we kindly ask you to take 5 minutes and answer the following questions.*

The following questions require general information from you.

- I am ...

  □ Working in this building

  □ Studying in this building

  □ Visiting the building

- What do you use the space of the restaurant for?

  □ Lunch

  □ Studying/Working

  □ Both, lunch and sometimes studying/working

  □ Other activities, namely _____

I am aware of sustainability issues	
I engage in environmentally friendly behaviour	
I am familiar with automated control in buildings	
I am familiar with lamps triggered by movement sensors	
I do not pay too much attention to technology	

1-totally disagree, 5 - totally agree

The following questions relate to the usefulness and effectiveness of au-
tomated lighting system that you might have experienced. For example,
when there is not enough natural light, the lamps can be switched on
by your movement in front of a sensor.

What do you think about the automated lighting system?	

1 - Not useful at all, 5 - Very useful, 6 - I do not know

The system saves energy	
The system considers natural light level	
The system considers people's presence	
The system does not cause distraction	
Switching lamps causes distraction	
Sensors detect my presence	
I have waved so as for a sensor to detect me	
I have got up and walked so as for a sensor to detect me	
I know that there are sensors in the restaurant	
The lamps stay on long enough after their switch	
I find the duration of lamps being switched on important	
For me, it is easy to use the lighting system	
I think that there is always enough light for me to perform my activity	
The system reacts immediately to changes	

1 - Not useful at all, 5 - Very useful, 6 - I do not know

What is your overall satisfaction with the system?	

1 - Totally dissatisfied , 5 - Totally satisfied, 6 - Does not apply

Appendix 5 - Evaluation sleep-mode software

A while ago you have indicated that you would like to be part of the testing the sleep-mode software for energy savings. The sleep-mode software test is part of the Sustainable Bernoulliborg project. We would like to know how you experienced having sleep mode enabled on your computer. Thank you for your time, your feedback is valuable to us.

Q1. I turn off the computer after leaving the office
□ *Always* □ *Very frequently* □ *Frequently* □ *Occasionally* □ *Rarely* □ *Very rarely* □ *Never* □ *I don't know*

Q2. Enabling sleep mode has made me more aware of how I use my computer
□ *Strongly agree* □ *Agree* □ *Neutral* □ *Disagree* □ *Strongly disagree* □ *I don't know*

Q3. Enabling sleep mode has made me put the computer to sleep more often
□ *Strongly agree* □ *Agree* □ *Neutral* □ *Disagree* □ *Strongly disagree* □ *I don't know*

Q4. Enabling sleep mode has made me turn the computer off more often
□ *Strongly agree* □ *Agree* □ *Neutral* □ *Disagree* □ *Strongly disagree* □ *I don't know*

Q5. On some occasions the computer entered sleep mode while I was actively using the computer
□ *Always* □ *Very frequently* □ *Frequently* □ *Occasionally* □ *Rarely* □ *Very rarely* □ *Never* □ *I don't know*

Q6. Having to wake up my computer from sleep mode disrupts my workflow
□ *Not much* □ *Little* □ *Somewhat* □ *Much* □ *A great deal* □ *I don't know*

Appendix 6 - Social acceptability evaluation (upon project completion)

As a part of the Sustainable Bernoulliborg project, several interventions in electricity consumption, water conservation and responsible waste recycling were performed in the Bernoulliborg. To evaluate acceptance of the interventions by you, the users and the buildings occupants, we would like to ask you to take part in a survey. This would give us new insights on improvements that will be used for purpose of research and publication, as well as advice for future interventions.

Displaying power consumption

The Energy Dashboard has been installed on several displays at the entrance of the building to provide feedback on the total electricity consumption of the building. This way, the occupants can see the real-time consumption of the building and decide to take saving actions to reduce the electricity consumption.

Q1. Seeing real-time and historic power consumption of the building raised my awareness about electricity used within our working environment.
    □ *Strongly agree* □ *Agree* □ *Neutral* □ *Disagree* □ *Strongly disagree* □ *I don't know*

Q2. Motivated by power consumption display, I performed energy saving actions at WORK more frequently.
    □ *Strongly agree* □ *Agree* □ *Neutral* □ *Disagree* □ *Strongly disagree* □ *I don't know*

Q3. If your answer to the previous question was affirmative, please state which energy saving actions you performed.
    ....

Q4. Motivated by the power consumption display, I performed energy saving actions at HOME more frequently.
    □ *Strongly agree* □ *Agree* □ *Neutral* □ *Disagree* □ *Strongly disagree* □ *I don't know*

Q5. In my opinion, displaying power consumption was an acceptable intervention.
    □ *Strongly agree* □ *Agree* □ *Neutral* □ *Disagree* □ *Strongly disagree* □ *I don't know*

Q6. I was satisfied with the quality of power consumption display.
    □ *Strongly agree* □ *Agree* □ *Neutral* □ *Disagree* □ *Strongly disagree* □ *I don't know*

Q7. If your answer to the previous question was negative, please state what aspects of the consumption display could be improved.

....

Adjustment of existing light/movement sensors

In several offices, consumption of office lights has been affected by adding a square sensor holder which enables better positioning of the movement sensor and more economic sensor settings. That gives better presence detection as well as the opportunity to reduce the time the lights are turned ON when not necessary to a minimum value (e.g., when employees leave office after finishing their work).

Q8. Adjustment of existing movement sensor increased my satisfaction with how the lights are controlled within my office.

□ *Strongly agree* □ *Agree* □ *Neutral* □ *Disagree* □ *Strongly disagree* □ *I don't know*

Q9. I was satisfied with the quality of the added sensor holder.

□ *Strongly agree* □ *Agree* □ *Neutral* □ *Disagree* □ *Strongly disagree* □ *I don't know*

Q10. In my opinion, existing movement sensors adjustment was an acceptable intervention.

□ *Strongly agree* □ *Agree* □ *Neutral* □ *Disagree* □ *Strongly disagree* □ *I don't know*

Water flow reduction

To save water, water flow reductors have been installed to all water faucets within the building. **Q11.** Adding water flow reductors on water faucets in the building did not decrease my satisfaction regarding water usage from faucets.

□ *Strongly agree* □ *Agree* □ *Neutral* □ *Disagree* □ *Strongly disagree* □ *I don't know*

Q12. Motivated by water flow reduction action, I performed water saving actions at WORK more frequently.

□ *Strongly agree* □ *Agree* □ *Neutral* □ *Disagree* □ *Strongly disagree* □ *I don't know*

Q13. If your answer for the previous question was affirmative, please state which water saving actions you performed.

....

Q14. Motivated by the water flow reduction action, I performed water saving actions at HOME more frequently.
   □ *Strongly agree* □ *Agree* □ *Neutral* □ *Disagree* □ *Strongly disagree* □ *I don't know*

Q15. In my opinion, the water flow reduction was an acceptable intervention.
   □ *Strongly agree* □ *Agree* □ *Neutral* □ *Disagree* □ *Strongly disagree* □ *I don't know*

Overall project

Q16. My overall satisfaction with the Sustainable Bernoulliborg project is:
   □ *Very satisfied* □ *Satisfied* □ *Neither satisfied nor dissatisfied* □ *Dissatisfied* □ *Very dissatisfied* □ *I don't know*

Q17. I think that the effects of this project could be increased if:
....

Q18. Other comments, questions or remarks
....

Bibliography

Abrahamse, W., Steg, L., Vlek, C. and Rothengatter, T.: 2005, A review of intervention studies aimed at household energy conservation, *Journal of Environmental Psychology* **25**(3), 273 – 291.

Abrahamse, W., Steg, L., Vlek, C. and Rothengatter, T.: 2007, The effect of tailored information, goal setting, and tailored feedback on household energy use, energy-related behaviors, and behavioral antecedents, *Journal of Environmental Psychology* **27**(4), 265 – 276.

Advantic Sys.: 2013, http://www.advanticsys.com/.

Agarwal, Y., Balaji, B., Gupta, R., Lyles, J., Wei, M. and Weng, T.: 2010, Occupancy-driven energy management for smart building automation, *Proceedings of the 2nd ACM Workshop on Embedded Sensing Systems for Energy-Efficiency in Building*, ACM, pp. 1–6.

Agile Manifesto: 2015, http://www.agilemanifesto.org.

Amit Dumbre, Sathya Priya Senthil, S. S. G.: 2011, Practicing agile software development on the windows azure platform, http://www.infosys.com/cloud/resource-center/Documents/practicing-agile-software-development.pdf.

AMQP Messaging System: 2013, http://www.amqp.org/.

Amstalden, R. W., Kost, M., Nathani, C. and Imboden, D. M.: 2007, Economic potential of energy-efficient retrofitting in the swiss residential building sector: the effects of policy instruments and energy price expectations, *Energy Policy* **35**(3), 1819–1829.

Armstrong, P., Agarwal, A., Bishop, A., Charbonneau, A., Desmarais, R. J., Fransham, K., Hill, N., Gable, I., Gaudet, S., Goliath, S., Impey, R., Leavett-Brown, C., Ouellete, J., Paterson, M., Pritchet, C., Penfold-Brown, D., Podaima, W., Schade, D. and Sobie, R. J.: 2010, Cloud scheduler: a resource manager for distributed compute clouds, *CoRR* **abs/1007.0050**.

Arshad, N., Heimbigner, D. and Wolf, A. L.: 2003, Deployment and dynamic reconfiguration planning for distributed software systems, *International Conference on Tools with Artificial Intelligence*, ICTAI, pp. 39–46.

Banfi, S., Farsi, M., Filippini, M. and Jakob, M.: 2008, Willingness to pay for energy-saving measures in residential buildings, *Energy economics* **30**(2), 503–516.

Berners-Lee, M.: 2010, *How Bad Are Bananas?: The Carbon Footprint of Everything*, Profile Books Ltd.

Boyano, A., Hernandez, P. and Wolf, O.: 2013, Energy demands and potential savings in european office buildings: Case studies based on energyplus simulations, *Energy and Buildings* **65**, 19–28.

Brambley, M. R.: 2005, *Advanced sensors and controls for building applications: Market assessment and potential R&D pathways*, Citeseer.

Bülow-Hübe, H.: 2008, Daylight in glazed office buildings.

Buyya, R., Yeo, C. S., Venugopal, S., Broberg, J. and Brandic, I.: 2009, Cloud computing and emerging it platforms: Vision, hype, and reality for delivering computing as the 5th utility, *Future Gener. Comput. Syst.* **25**, 599–616.

Caruso, M., Mecella, M., Baldoni, R., Querzoni, L. and Cerocchi, A.: 2013, User profiling and micro-accounting for smart energy management, *The 11th ACM Conference on Embedded Network Sensor Systems, SenSys '13, Roma, Italy, November 11-15, 2013*, pp. 42:1–42:2.

Caruso, M., Mecella, M., Cerocchi, A., Forte, V., Querzoni, L. and Baldoni, R.: 2014, Energy management in smart spaces through the oplatform, *2014 International Conference on Intelligent Networking and Collaborative Systems, Salerno, Italy, September 10-12, 2014*, pp. 563–568.

Cato, M.: 2009, *Green Economics: An Introduction to Theory, Policy and Practice*, Earthscan.

Chaves, C., Batista, D. and da Fonseca, N.: 2010, Scheduling grid applications on clouds, *Global Telecommunications Conference (GLOBECOM 2010), 2010 IEEE*, pp. 1 –5.

Chen, B., Potts, C. and Woeginger, G.: 1999, *Handbook of Combinatorial Optimization*, Springer US, chapter A Review of Machine Scheduling: Complexity, Algorithms and Approximability, pp. 1493–1641.

CollabNet, I.: 2011, Reinforcing agile software development in the cloud, http://www.open.collab.net/media/pdfs/CollabNet

Costanzo, M., Archer, D., Aronson, E. and Pettigrew, T.: 1986, Energy conservation behavior: The difficult path from information to action., *American psychologist* **41**(5), 521.

Darby, S. et al.: 2006, The effectiveness of feedback on energy consumption, *A Review for DEFRA of the Literature on Metering, Billing and direct Displays* **486**, 2006.

Davis, F.: 1986, A technology acceptance model for empirically testing new end-user information system: Theory and results.

Davis, F. D., Bagozzi, R. P. and Warshaw, P. R.: 1989, User acceptance of computer technology: A comparison of two theoretical models, *Management Science* **35**(8), 982–1003.

Degeler, V.: 2014, *Dynamic Rule-Based Reasoning in Smart Environments*, PhD thesis, University of Groningen.

Degeler, V. and Lazovik, A.: 2013, Architecture pattern for context-aware smart environments, *in* P. H. D. Riboni, B. Guo (ed.), *Creating Personal, Social and Urban Awareness through Pervasive Computing*, IGI Global.

Dekker, M. and Brandsma, J.: 2015, An energy competition dashboard, *Bachelor thesis*, University of Groningen.

Deng, Y., Li, Z. and Quigley, J. M.: 2012, Economic returns to energy-efficient investments in the housing market: evidence from singapore, *Regional Science and Urban Economics* **42**(3), 506–515.

Di Cosmo, R., Zacchiroli, S. and Zavattaro, G.: 2012, Towards a formal component model for the cloud, *International Conference on Software Engineering and Formal Methods*, SEFM, pp. 156–171.

Dragoicea, M., Bucur, L. and Patrascu, M.: 2013, A service oriented simulation architecture for intelligent building management, *Proceedings of the 4th International Conference on Exploring Service Science*, Vol. LNBIP 143, Springer, pp. 14–28.

drs. T.A. van den Broek, dr. M.L. Ehrenhard, Langley, D. and Groen, P. A.: 2012, Dotcauses for sustainability: combining activism and entrepreneurship, *Journal of public affairs* **12**(3), 214 – 223.

Dubois, M.-C. and Blomsterberg, Å.: 2011, Energy saving potential and strategies for electric lighting in future North European, low energy office buildings: A literature review, *Energy and Buildings* **43**(10), 2572–2582.

Erol, K., Hendler, J. and Nau, D. S.: 1994, Htn planning: Complexity and expressivity, *National Conference on Artificial Intelligence*, AAAI'94, AAAI, pp. 1123–1128.

Fdez-Olivares, J., Castillo, L., García-Pérez, O. and Palao, F.: 2006, Bringing users and planning technology together. Experiences in SIADEX, *International Conference on Automated Planning and Scheduling*, ICAPS, pp. 11–20.

Ge, Y. and Wei, G.: 2010, Ga-based task scheduler for the cloud computing systems, *Web Information Systems and Mining (WISM), 2010 International Conference on*, Vol. 2, pp. 181 –186.

Georgievski, I.: 2013, HPDL: Hierarchical Planning Definition Language, *JBI Preprint 2013-12-3*, Uni. of Groningen.

Georgievski, I. and Aiello, M.: 2012, An Overview of Hierarchical Task Network Planning, *JBI Preprint 2012-12-5*, Uni. of Groningen.

Georgievski, I. and Aiello, M.: 2015, Htn planning: Overview, comparison, and beyond, *Artificial Intelligence* **222**(0), 124–156.

Georgievski, I. and Lazovik, A.: 2014, Utility-based htn planning, *European Conference on Artificial Intelligence*, p. 1013?1014.

Georgievski, I., Nguyen, T. A. and Aiello, M.: 2013, Combining activity recognition and ai planning for energy-saving offices, *10th Int'l. Conf. on Ubiquitous Intelligence and Computing/Autonomic and Trusted Computing (UIC/ATC)*, pp. 238–245.

Gilkey, H. T.: 1959, New air heating methods, *New methods of heating buildings: a research correlation conference conducted by the Building Research Institute, Division of Engineering and Industrial Research, as one of the programs of the BRI fall conferences*, National Research Council (U.S.). Building Research Institute, Washington, p. p. 60. OCLC 184031.

Groen, P. A. and Walsh, P. S.: 2013, Introduction to the field of emerging technology management, *Creativity and innovation management* **22**(1), 1 – 5.

Harrer, S., Nizamic, F., Wirtz, G. and Lazovik, A.: 2014, Towards a Robustness Evaluation Framework for BPEL Engines, *Proceedings of the 7th IEEE International Conference on Service-Oriented Computing and Applications (SOCA'14)*, IEEE, Matsue, Japan, pp. 199–206.

HermiT Reasoner: 2013, http://hermit-reasoner.com/.

Hevner, A. R., March, S. T., Park, J. and Ram, S.: 2004, Design science in information systems research, *MIS quarterly* **28**(1), 75–105.

Hoeksema, T. and Medema, M.: 2015, Sleepy for linux - power management framework for workstations, *Bachelor thesis*, University of Groningen.

Horsley, A., France, C. and Quatermass, B.: 2003, Delivering energy efficient buildings: a design procedure to demonstrate environmental and economic benefits, *Construction Management and Economics* **21**(4), 345–356.

Howe, J. C.: 2010, Overview of green buildings, *National Wetlands Newsletter* **33**(1).

Idsardi, J.: 2014, Recycling data for everyone.

Jans, J.: 2015, Building maintenance application system.

Jilings, M. and Heitmeijer, S.: 2013, Office energy saving potential through component based automation, a design and implementation, *Master thesis*, University of Groningen.

Kaldeli, E., Lazovik, A. and Aiello, M.: 2011, Continual planning with sensing for Web service composition, *AAAI Conference on Artificial Intelligence*, AAAI, pp. 1198–1203.

Kapsalaki, M., Leal, V. and Santamouris, M.: 2012, A methodology for economic efficient design of net zero energy buildings, *Energy and Buildings* **55**, 765–778.

Kawamoto, K., Shimoda, Y. and Mizuno, M.: 2004, Energy saving potential of office equipment power management, *Energy and Buildings* **36**(9), 915–923.

Ken Peffers, Tuure Tuunanen, M. A. R. S. C.: 2007, A design science research methodology for information systems research, *Journal of management information systems* **24**(3), 45–77.

Kim, S., Kim, Y., Song, N. and Kim, C.: 2010, Adaptable scheduling schemes for scientific applications on science cloud, *Cluster Computing Workshops and Posters (CLUSTER WORKSHOPS), 2010 IEEE International Conference on*, pp. 1 –3.

Lakshman, A. and Malik, P.: 2010, Cassandra: a decentralized structured storage system, *Operating Systems Review* **44**(2), 35–40.

Larsson, L., Henriksson, D. and Elmroth, E.: 2011, Scheduling and monitoring of internally structured services in cloud federations, *Computers and Communications (ISCC), 2011 IEEE Symposium on*, pp. 173 –178.

Lascu, T. A., Mauro, J. and Zavattaro, G.: 2013, A planning tool supporting the deployment of cloud applications, *International Conference on Tools with Artificial Intelligence*, ICTAI, pp. 213–220.

Lindén, A.-L., Carlsson-Kanyama, A. and Eriksson, B.: 2006, Efficient and inefficient aspects of residential energy behaviour: What are the policy instruments for change?, *Energy policy* **34**(14), 1918–1927.

Lu, X. and Gu, Z.: 2011, A load-adapative cloud resource scheduling model based on ant colony algorithm, *Cloud Computing and Intelligence Systems (CCIS), 2011 IEEE International Conference on*, pp. 296 –300.

MacKay, D. J. C.: 2008, *Sustainable Energy – without the hot air*, UIT Cambridge. ISBN 978-0-9544529-3-3. Available free online from www.withouthotair.com.

Martin, D., Burstein, M., Mcdermott, D., Mcilraith, S., Paolucci, M., Sycara, K., Mcguinness, D. L., Sirin, E. and Srinivasan, N.: 2007, Bringing semantics to Web services with OWL-S, *World Wide Web* **10**(3), 243–277.

Mecella, M. and Baldoni, R.: 2011, Smart homes for all: Simplifying lives with service composition over embedded devices, *ERCIM News* **2011**(87).

Meiboom, M.: 2013, Mobile visualisation of energy consumption data for improving awareness amongst building occupants.

Musters, B., Wiersma, N. and Kluiter, N.: 2014, Monitoring water usage of the bernoulliborg.

Nguyen, T. A. and Aiello, M.: 2013, Energy intelligent buildings based on user activity: A survey, *Energy and Buildings* **56**(0), 244–257.

Nguyen, T. A., Degeler, V., Contarino, R., Lazovik, A., Bucur, D. and Aiello, M.: 2013, Towards context consistency in a rule-based activity recognition architecture, *10th Int'l. Conf. on Ubiquitous Intelligence and Computing/Autonomic and Trusted Computing (UIC/ATC)*, pp. 625–630.

Nguyen, T. A., Raspitzu, A. and Aiello, M.: 2013, Ontology-based office activity recognition with applications for energy savings, *Journal of Ambient Intelligence and Humanized Computing* pp. 1–15.

Nikolaidis, Y., Pilavachi, P. A. and Chletsis, A.: 2009, Economic evaluation of energy saving measures in a common type of greek building, *Applied Energy* **86**(12), 2550–2559.

Nizamic, F.: 2013, Testing of distributed service-oriented systems, *ICSOC*, Vol. 8377 of *Lecture Notes in Computer Science*, Springer Berlin Heidelberg, Berlin, Germany.

Nizamic, F., Degeler, V., Groenboom, R. and Lazovik, A.: 2012, Policy-based scheduling of cloud services, *Scalable Computing: Practice and Experience* **13**(3), 187–199.

Nizamic, F., Groenboom, R. and Lazovik, A.: 2011, Testing for highly distributed service-oriented systems using virtual environments, *Postproceedings of 17th Dutch Testing Day*.

Nizamic, F., Nguyen, T. A., Lazovik, A. and Aiello, M.: 2014, Greenmind - an architecture and realization for energy smart buildings, *Proceedings of the 2nd International Conference on ICT for Sustainability (ICT4S 2014)*, Stockholm, Sweden. doi:10.2991/ict4s-14.2014.3.

Oikonomou, V., Becchis, F., Steg, L. and Russolillo, D.: 2009, Energy saving and energy efficiency concepts for policy making, *Energy Policy* **37**(11), 4787–4796.

Pernici, B., Aiello, M., vom Brocke, J., Donnellan, B., Gelenbe, E. and Kretsis, M.: 2012, What is can do for environmental sustainability, *Communications of the Association for Information Systems* **30**(18).

Plugwise: 2013, http://www.plugwise.com/.

Poortinga, W., Steg, L., Vlek, C. and Wiersma, G.: 2003, Household preferences for energy-saving measures: A conjoint analysis, *Journal of Economic Psychology* **24**(1), 49–64.

Protégé: 2013, http://protege.stanford.edu/.

Riabov, A. and Liu, Z.: 2005, Planning for stream processing systems, *National Conference on Artificial Intelligence*, AAAI, pp. 1205–1210.

Roth, K. W., Westphalen, D., Dieckmann, J., Hamilton, S. D. and Goetzler, W.: 2002, *Energy Consumption Characteristics of Commercial Building HVAC Systems: Volume III, Energy Savings Potential*, National Technical Information Service (NTIS), U.S. Department of Commerce.

Setz, B.: 2015, Machine learning-assisted optimisation of sensor settings based on user behaviour, *Master thesis*, University of Groningen.

Setz, B. and Pul, R.: 2013, Lazy sleep, *Internship report*, University of Groningen.

Sirin, E., Parsia, B., Wu, D., Hendler, J. and Nau, D. S.: 2004, HTN planning for Web service composition using SHOP2, *Web Semantic* **1**, 377–396.

Sohrabi, S., Prokoshyna, N. and Mcilraith, S. A.: 2006, Web service composition via generic procedures and customizing user preferences, *International Semantic Web Conference*, ISWC'06, pp. 597–611.

Sohrabi, S., Udrea, O. and Riabov, A.: 2013, HTN planning for the composition of stream processing applications, *International Conference on Automated Planning and Scheduling*, ICAPS, pp. 443–451.

Sotiriadis, S., Bessis, N. and Antonopoulos, N.: 2011, Towards inter-cloud schedulers: A survey of meta-scheduling approaches, *P2P, Parallel, Grid, Cloud and Internet Computing (3PGCIC), 2011 International Conference on*, pp. 59 –66.

Steg, L.: 2008, Promoting household energy conservation, *Energy Policy* **36**(12), 4449–4453.

Tanenbaum, A. S. and Steen, M. v.: 2006, *Distributed Systems: Principles and Paradigms (2Nd Edition)*, Prentice-Hall, Inc., Upper Saddle River, NJ, USA.

The Cassandra database: 2014, http://cassandra.apache.org/.

The Neo4j database: 2014, http://www.neo4j.org/.

Uittenbogaard, B., Broens, L. and Groen, A. J.: 2005, Towards a guideline for design of a corporate entrepreneurship function for business development in medium-sized technology-based companies, *Creativity and Innovation Management* **14**(3), 258–271.

UNEP: 2009, Buildings and climate change.

Vaquero, L. M., Rodero-Merino, L., Caceres, J. and Lindner, M.: 2008, A break in the clouds: towards a cloud definition, *SIGCOMM Comput. Commun. Rev.* **39**, 50–55.

Veenstra, A., Stollenga, M. and Andringa, T.: 2008, Herkennen van aan- en uitschakelmomenten in energiemeterdata, *Report*, University of Groningen.

Venkatesh, V.: 2000, Determinants of perceived ease of use: Integrating control, intrinsic motivation, and emotion into the technology acceptance model, *Info. Sys. Research* **11**(4), 342–365.

Venkatesh, V. and Davis, F. D.: 2000, A theoretical extension of the technology acceptance model: Four longitudinal field studies, *Management Science* **46**(2), 186–204.

Venkatesh, V., Morris, M. G., Davis, G. B. and Davis, F. D.: 2003, User acceptance of information technology: Toward a unified view, *MIS Quarterly* **27**(3), pp. 425–478.

Videla, A. and Williams, J. J.: 2012, *RabbitMQ in Action*, Manning.

Webber, C. A., Roberson, J. A., McWhinney, M. C., Brown, R. E., Pinckard, M. J. and Busch, J. F.: 2006, After-hours power status of office equipment in the usa, *Energy* **31**(14), 2823–2838.

Wilhite, H., Nakagami, H., Masuda, T., Yamaga, Y. and Haneda, H.: 1996, A cross-cultural analysis of household energy use behaviour in japan and norway, *Energy Policy* **24**(9), 795–803.

Yumatov, S.: 2010, Migration of smog to amazon web services, *Report*, University of Groningen.

Zigbee: 2015, http://www.zigbee.org.

Samenvatting

Bestaande gebouwen zijn verantwoordelijk voor meer dan 40% van het totale energieverbruik wereldwijd. Er wordt geschat dat dergelijke gebouwen verantwoordelijk zijn voor een derde van de totale broeikasgasemissies in de wereld, dit komt voornamelijk door het gebruik van fossiele brandstoffen gedurende de tijd dat de gebouwen operationeel zijn. Uit onderzoek blijkt dat gemiddeld 30% van de gebruikte energie in niet-residentile gebouwen, zoals kantoor- of universiteitsgebouwen, verspild wordt. Het is daarom van cruciaal belang dat het verbruik van energie en andere middelen in niet-residentile gebouwen wordt teruggebracht door verbeteringen in het operationele proces.

Tegenwoordig worden veel niet-residentile gebouwen bestuurd door gebouwbeheersystemen (GBS). Een GBS is een computersysteem om de mechanische en elektrische systemen van een gebouw aan te sturen en te monitoren. Bestaande GBS-en zijn echter niet in staat om zowel het onnodige energieverbruik terug te brengen als het comfortniveau van de gebruiker te behouden. Een van de redenen hiervoor is dat GBS-en over het algemeen vooraf ingeplande handelingen uitvoeren en daardoor niet om kunnen gaan met de veranderingen in het gebouw veroorzaakt door de wisselwerking tussen de gebruiker en de omgeving.

We ontwerpen, implementeren, optimaliseren en evalueren een slim energiesysteem dat rekening houdt met het gedrag van de gebruikers in een gebouw door automatisch het energieverbruik te sturen alsmede het aansturen van apparatuur in een daadwerkelijk operationele omgeving. Hiermee worden een aantal belangrijke technologische bijdragen geleverd: ten eerste verbeteren we het beheer van de software-infrastructuur door gebruik te maken van kleine, onafhankelijke services ('micro-services'). Door het gebruik van deze infrastuctuur wordt het systeem schaalbaar en vereenvoudigt het de bouw, implementatie en het uitvoeren van gedistribueerde systemen. Ten tweede stroomlijnen we de software-architectuur door historische data een belangrijke rol te laten spelen. Ten derde richten wij ons op 'Graphical User Interfaces (GUIs), aangezien de grootste winst te behalen is bij de interactie tussen mens en machine. GUIs worden gebruikt om gebruikers te voorzien van belangrijke informatie betreffende hun energieverbruik, om mileubewustzijn te verhogen en om ze actief dan wel passief te betrekken in het verminderen van het energieverbruik.

Naast ons belangrijkste doel, het verminderen van het energieverbruik in gebouwen, onderzoeken wij ook hoe we dezelfde principes op vergelijkbare situaties kunnen toepassen, zoals bijvoorbeeld het waterverbruik verminderen en het afvalverwerkingsproces optimaliseren. Bovendien maken we gebruik van op cloud-technologie gebaseerde services bij het implementeren van het voorgestelde systeem, met als doel om het gebruik van beschikbare computercapaciteit te optimaliseren. Dit zal leiden tot verdere energiebesparingen. Cloud-systemen bieden schaalbaarheid doordat een enkele installatie gebruikt kan worden door meerdere partijen. Om de implementatie van de verschillende soorten services te optimaliseren maken we gebruik van scheduling- en planning technieken. Daarnaast evalueren we de prestaties van deze technieken door middel van experimenten.

Voor dit onderzoek is de Bernouliborg, een modern universiteitsgebouw, gebruikt als case study. In dit gebouw hebben wij een prototype geplaatst om te laten zien hoe een gentegreerd systeem verwezenlijkt kan worden in een kantooromgeving. Het gebouw is gebruikt als evaluatieplatform voor ons onderzoek en voor het uitvoeren van benchmarks op het gebied van mileu- en sociale overwegingen. Daarnaast is het gebouw ook gebruikt voor de beoordeling van de economische besparingen.

We laten zien dat de behaalde mileubesparingen van de gemplementeerde oplossingen variren van 9,5% voor het aansturen van de slaapstand van computers, tot 80% voor het automatisch aansturen van verlichtingsinstallaties. De kostenbesparing laat zien dat de door ons ontwikkelde systemen zichzelf terugverdienen in ongeveer twee jaar bij het verminderen van de waterconsumptie en binnen 7,8 jaar voor het automatisch in en uitschakelen van de verlichting. De terugverdientijd is afhankelijk van de totale investering en van de jaarlijkse besparingen die gerealiseerd worden door efficinte oplossingen. Een evaluatie van de gebruikersacceptatie laat zien dat een meerderheid van de deelnemers de systemen als zeer acceptabel en nuttig ondervinden.

www.ingramcontent.com/pod-product-compliance
Lightning Source LLC
Chambersburg PA
CBHW051409200326
41520CB00023B/7174